HARPERCOLLINS COLLEGE OUTLINE

Finite Mathematics with Calculus

Joan Dykes, Ph.D.
Edison Community College

Ronald Smith, Ph.D.
Edison Community College

HarperPerennial
A Division of HarperCollins*Publishers*

 For information address HarperCollins Publishers, Inc., 10 East 53rd Street, New York, NY 10022.

FIRST HARPERPERENNIAL EDITION

An American BookWorks Corporation Production

Project Manager: William Hamill
Editor: Mark N. Weinfeld

Library of Congress Cataloging-in-Publication Data

Dykes, Joan, 1951–
Finite mathematics with calculus / Joan Dykes, Ronald Smith.
p. cm. — (HarperCollins college outline)
Includes index.
ISBN: 0-06-467164-X (pbk.)
1. Mathematics. 2. Calculus. I. Smith, Ron, 1949–
II. Title. III. Series.
QA37.2. D95 1993 92–54680

94 95 96 97 ABW/RRD 10 9 8 7 6 5 4 3 2 1

Contents

Preface

This book is provided as a supplement to a standard college finite mathematics textbook that includes coverage of some calculus topics. Although a knowledge of intermediate algebra is assumed, as many steps as possible are provided to help the reader follow the logic involved in solving the various problems encountered in a course of this nature. Theorems are stated without proof and are often restated in words or symbols more easily understood by our own students. A set of exercises and answers appear at the end of each chapter to allow the reader to practice and receive immediate feedback. After working the exercises we have provided, use your own textbook exercises to develop the skill and speed to perform well in your course. Keep pencil and paper handy—reading and working through this book will help you succeed in finite mathematics and calculus.

Joan Dykes
Ronald Smith

1

Lines

This chapter covers the techniques used in graphing line functions in two dimensions. Graphs of linear inequalities will be demonstrated. The concepts shown will be in processes involving systems of linear equations and business applications involving those systems of equations.

1.1 EQUATIONS OF LINES

All of the functions of the form $f(x) = mx + b$ are functions that represent a line. These functions will show a straight line when graphed. In this format, m is the slope of the line, and b is the y-intercept of the line. There are various aspects of these straight line functions that are studied in depth. These aspects will be discussed in this section.

Slope

The "steepness" of a line is called the *slope* of the line. The slope of the line is the ratio of the vertical movement along the line (*rise*) to the horizontal movement along the line (*run*).

$$\textbf{slope} = \frac{\text{rise}}{\text{run}}$$

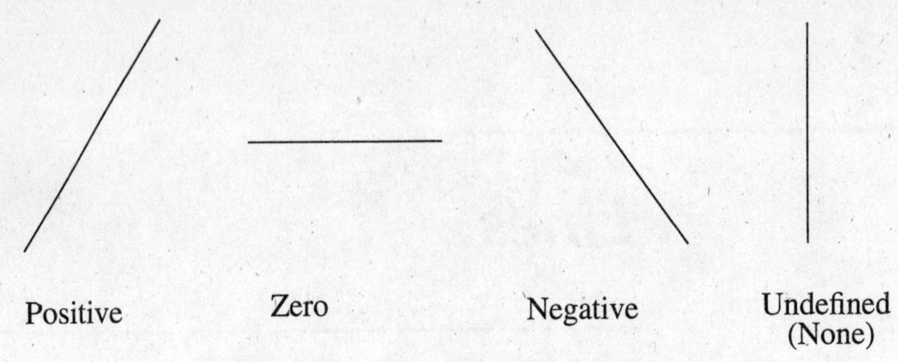

Figure 1.1 The Various Types of Slope

Figure 1.1 shows how the slope of the line relates to its steepness. The less steep the line is, the smaller the absolute value of the slope. This figure shows that a line that is **horizontal** has a **slope of 0**. As the slant of the line becomes steeper from left to right, the slope becomes a larger positive number.

Although the nuances of our language often allow the substitution of "zero" for "no," this cannot be the case when we talk about slopes. It is important to understand that a vertical line doesn't have a slope and a horizontal line has a slope equal to the number 0.

Since the rise on a graph is the difference between ***y***-values, and the run on the graph is the difference between ***x***-values, a slope can be determined from two points (x_1, y_1) and (x_2, y_2) on a graph by the following formula. The letter ***m*** is the mathematical designation for **slope**.

$$m = \frac{y_2 - y_1}{x_2 - x_1}$$

The line that connects point A, (3, 2) , and point B, (−1, 4), has a slope defined by:

$$m = \frac{4-2}{-1-3} = \frac{2}{-4} = -\frac{1}{2}$$

Note: It does not matter which point is labeled point A or which is labeled point B; the slope that is computed will be the same.

EXAMPLE 1.1

Find the slope of the line passing through the following points. It does not matter which point is labeled (x_1, y_1) and which is labeled (x_2, y_2):

(a) $(2, 1)$ and $(7, 1)$
(b) $(2, 0)$ and $(-3, -1)$
(c) $(1, -2)$ and $(1, 3)$

SOLUTION 1.1

(a) $m = \dfrac{y_2 - y_1}{x_2 - x_1} = \dfrac{1 - 1}{7 - 2} = \dfrac{0}{5} = 0$

(b) $m = \dfrac{y_2 - y_1}{x_2 - x_1} = \dfrac{0 - (-1)}{2 - (-3)} = \dfrac{1}{5}$

(c) $m = \dfrac{y_2 - y_1}{x_2 - x_1} = \dfrac{3 - (-2)}{1 - 1} = \dfrac{5}{0} =$ undefined

Division by 0 is undefined.

Two formulas are often used to convert information about a line into an equation for that line. These two formulas are

slope-intercept form $\quad y = mx + b$
where: m is the slope
b is the y-intercept (the point where the line crosses the y-axis)

point-slope form $\quad y - y_1 = m(x - x_1)$
where: m is the slope
(x_1, y_1) are the coordinates of one point on the line
Note: x and y (without subscripts) are not replaced by numbers.

Some additional information will be helpful as we seek to find equations of lines:

(a) The slopes of two **parallel lines** with slopes of m_1 and m_2, respectively, are equal. That is, $m_1 = m_2$.
(b) The slopes of two perpendicular lines are the negative reciprocals of

each other. That is, $m_1 = \frac{-1}{m_2}$.

(c) The coordinates of the origin are $(0, 0)$.

(d) The point at which the line crosses the x-axis is called the x-intercept. Its coordinates are (x-value, 0).

(e) The points at which the line crosses the y-axis is called the y-intercept. The coordinates for the y-intercept are (0, y-value).

Armed with the above information and the two formulas for developing the line equations, we can now find the equation of a line given various information.

EXAMPLE 1.2

(a) Determine the equation of the line passing through the points $(2, 1)$ and $(-1, 3)$.

(b) Find the equation of the line with slope of $\frac{3}{2}$ and a y-intercept of –2.

(c) Write the equation of a line with x-intercept of 2 and passing through $(1, 5)$.

(d) Find the equation of a line passing through the origin and crossing $(-7, 1)$.

(e) Find the equation of a line with x-intercept of –5 and y-intercept of –2.

(f) Formulate the equation of the line parallel to the line $2x - 4y = 3$ and passing through $(8, 0)$.

(g) Determine the equation of a line perpendicular to $y = -\frac{1}{5}x + 1$ and passing through $(0, -1)$.

(h) Find the equation of the horizontal line passing through $(-2, 3)$.

(i) Write the equation of the vertical line crossing the x-axis at –7.

SOLUTION 1.2

(a) With two points given, we can first find the slope and then use one of the points and the slope in the point-slope formula.

$(2, 1) \quad (-1, 3)$ Two points are given.

$$m = \frac{y_2 - y_1}{x_2 - x_1} = \frac{3 - 1}{-1 - 2} = \frac{2}{-3} = -\frac{2}{3}$$ Find the slope.

$m = \frac{-2}{3} \qquad (2, 1) = (x_1, y_1)$ Select either one of the points and use the slope found earlier.

$y - y_1 = m(x - x_1)$ We now need the point-slope formula.

$y - 1 = -\frac{2}{3}(x - 2)$ Substitute for slope and point.

$(3)\,(y-1) = (3)\left(-\dfrac{2}{3}\right)(x-2)$	Multiply both sides of the equation by 3.
$3\,(y-1) = -2\,(x-2)$	
$3y-3 = -2x+4$	Distribute to remove parentheses.
$\begin{array}{rcl} 3y-3 &=& -2x+4 \\ +3 & & +3 \\ \hline \end{array}$	Add 3 to both sides.
$\begin{array}{rcl} 3y &=& -2x+7 \\ +2x & & +2x \\ \hline \end{array}$	Add $2x$ to both sides.
$2x+3y = 7$	This completes the equation and isolates the constant (7).

(b) Knowing the slope $(m = 3/2)$ and the y-intercept $(b = -2)$, this problem is tailor-made for the slope-intercept formula.

$y = mx+b$	Write the slope-intercept form of the equation.
$y = \dfrac{3}{2}x + (-2)$	Substitute slope and intercept.
$y = \dfrac{3}{2}x-2$	You could leave the equation in this form.
$(2)\,y = 2\left(\dfrac{3}{2}x-2\right)$	Or you could change it into the standard form by multiplying by 2.
$2y = 2\left(\dfrac{3}{2}x\right)-2\,(2)$	Don't forget to distribute the 2 on the right through both terms.
$2y = 3x-4$	
$-3x+2y = -4$	Isolate the constant term.
or	
$3x-2y = 4$	

(c) In this case the equation will be derived from two points (the x-intercept and the point $(1, 5)$).

An x-intercept of 2 means the line contains the point $(2, 0)$. Thus, two points this line passes through are $(2, 0)$ and $(1, 5)$.

$$m = \frac{y_2-y_1}{x_2-x_1} = \frac{5-0}{1-2} = \frac{5}{-1} = -5$$ Find the slope.

$y - y_1 = m(x - x_1)$	We now need the point-slope formula.
$m = -5 \qquad (x_1, y_1) = (2, 0)$	
$y - 0 = -5(x - 2)$	Substitute into the formula.
$y = -5(x - 2)$	
$y = -5x + 10$	Distribute.
$5x + y = 10$	Isolate the constant term.

(d) This equation will also be built from two points. We will use the origin $(0, 0)$ and the point $(-7, 1)$.

$m = \dfrac{y_2 - y_1}{x_2 - x_1} = \dfrac{1 - 0}{-7 - 0} = \dfrac{-1}{7}$	Find the slope.
$y - y_1 = m(x - x_1)$	We now need the point-slope formula.
$y - 0 = \dfrac{-1}{7}(x - 0)$	Substitute into point-slope formula.
$y = \dfrac{-1}{7}x$	Simplify the equation.
$7y = 7\left(\dfrac{-1}{7}x\right)$	Multiply both sides of the equation by the denominator.
$7y = -x$	
$7y + x = -x + x$	Add x to both sides of the equation.
$x + 7y = 0$	

Note: The standard form of any line passing through the origin will always be an equation of the form $ax + by = 0$.

(e) This problem has two intercepts, which yield two points. The points are $(-5, 0)$ from the x-intercept of –5 and $(0, -2)$ from the y-intercept of –2.

$m = \dfrac{y_2 - y_1}{x_2 - x_1} = \dfrac{-2 - 0}{0 - (-5)} = \dfrac{-2}{5}$	Find the slope.
$m = \dfrac{-2}{5} \qquad b = -2$	We could use the point-slope form, but the slope-intercept is easier.

$y = mx + b$ — Write the slope-intercept form of the equation.

$y = \frac{-2}{5}x - 2$ — Substitute into the equation.

$(5)y = (5)\left(\frac{-2}{5}x - 2\right)$ — Multiply both sides of the equation by the denominator.

$5y = 5\left(\frac{-2}{5}x\right) - 5(2)$ — Distribute.

$5y = -2x - 10$

$5y + 2x = -2x + 2x - 10$ — Add $2x$ to both sides.

$2x + 5y = -10$ — Isolate the constant term.

(f) We have a slope "hidden" in the line that is given as parallel to the line we are to find and we are given a point (8, 0).

$2x - 4y = 3$ — Solve for y in terms of x to find the slope of the given line.

$-4y = -2x + 3$

$\frac{-4y}{-4} = \frac{-2x}{-4} + \frac{3}{-4}$ — Divide both sides by -4.

$y = \frac{1}{2}x - \frac{3}{4}$ — The equation is now in slope-intercept form. We will use only the slope and disregard the rest of the linear equation.

$(x_1, y_1) = (8, 0) \quad m = \frac{1}{2}$

$y - y_1 = m(x - x_1)$ — Use the point-slope form.

$y - 0 = \frac{1}{2}(x - 8)$ — Substitute for the slope and the point.

$2(y - 0) = 2\left(\frac{1}{2}\right)(x - 8)$ — Multiply both sides by 2.

$-x + 2y = -8$ — Subtract x from both sides.

or

$x - 2y = 8$

(g) The slope of the desired line is obtained from $y = -\frac{1}{5}x+1$. The slope of $y = -\frac{1}{5}x+1$ is the coefficient of x which is $-\frac{1}{5}$. Perpendicular lines have slopes that are negative reciprocals of each other. We find the new slope by taking $-\frac{1}{5}$ and "flipping" it to get $\frac{5}{-1} = -5$. Next, we change the sign to get 5. The slope of the line perpendicular to the given line is 5.

$(x_1, y_1) = (0, -1) \quad m = 5$	
$y - y_1 = m(x - x_1)$	Use the point-slope form.
$y - (-1) = 5(x - 0)$	Substitute for slope and point.
$y + 1 = 5x$	Simplify.
$y + 1 - 1 = 5x - 1$	Subtract 1 from both sides.
$y = 5x - 1$	

(h) The equation of a horizontal line through (x-value, y-value) is y = y-value. Since we were given (–2, 3), the equation must be $y = 3$.

(i) The equation of a vertical line through (x-value, y-value) is x = x-value. Since we were given (–7, 0), the equation must be $x = -7$.

1.2 SOLVING INEQUALITIES

Rules for Inequalities

While linear equations with one unknown have only one solution, linear inequalities usually have many (or a range) of solutions.

Symbol	Meaning
$x > y$	x is greater than y
$x < y$	x is less than y
$x \geq y$	x is greater than or equal to y
$x \leq y$	x is less than or equal to y

Table 1.1 Inequality Symbols

Rule	Format
addition	$a + c > b + c$
subtraction	$a - c > b - c$
multiplication	$a \cdot c > b \cdot c; \quad c > 0$
division	$a \div c > b \div c; \quad c > 0$
negation	$-a < -b$

Table 1.2 Inequality Rules For $a > b$

In Table 1.2, you might notice that a negation of the inequality reverses the direction (sometimes referred to as "the sense") of the inequality. That is, the > will reverse to < if the inequality is divided or multiplied by a negative number. Addition and subtraction do not affect the direction of the inequality.

EXAMPLE 1.3

Solve the following inequalities:

(a) $x + 5 > 7$

(b) $2x - 3 < 5$

(c) $4 - 5x > 24$

(d) $6 - \frac{x}{3} < 18$

SOLUTION 1.3

(a) $x + 5 > 7$ — Copy the given inequality.

$$\begin{array}{rcr} x + 5 & > & 7 \\ \underline{-5} & & \underline{-5} \end{array}$$

Subtract 5 from each side of the inequality.

$$x > 2$$

Solution set is all real numbers greater than 2.

(b) $2x - 3 < 5$ — Copy the given inequality.

$$\begin{array}{rcr} 2x - 3 & < & 5 \\ \underline{+3} & & \underline{+3} \\ 2x & < & 8 \end{array}$$

Add 3 to both sides of the inequality.

$$\frac{2x}{2} < \frac{8}{2}$$

Divide both sides by 2.

$$x < 4$$

The solution set includes all real numbers less than 4. Check by substituting numbers less than 4 into the original inequality.

(c) $4 - 5x > 24$

Copy the given inequality.

$$\begin{array}{r} 4 - 5x > 24 \\ \underline{-4 \qquad -4} \\ -5x > 20 \end{array}$$

Subtract 4 from each side of the inequality.

$$\frac{-5x}{-5} < \frac{20}{-5}$$

We divide by –5. Remember that when we divide by a negative number, we reverse the direction of the inequality.

$$x < -4$$

The direction of the inequality has been reversed. The solution set includes all numbers less than –4. Check by substituting numbers less than –4 into the original inequality.

(d) $6 - \frac{x}{3} < 18$

Copy the given inequality.

$$6 - \frac{x}{3} < 18$$

Subtract 6 from each side of the inequality.

$$-\frac{x}{3} < 12$$

$$-\frac{3}{1} \cdot \frac{-x}{3} > (-3)\,12$$

Multiply both sides of the inequality by (–3).

$$x > -36$$

The direction of the inequality is reversed.

$$x > -36$$

The solution set is any number greater than –36. Check by placing any number larger than –36 in the original inequality.

1.3 GRAPHING POINTS AND LINES

Graphing a Point

Each point drawn in two dimensions has an address to specify its location on the two-dimensional plane. Each address is called an *ordered pair*. The ordered pair is specified, in general, as (x, y), where x and y are real numbers. The x-value represents the distance of the point from the origin along the x-axis and the y-value represents the distance of the point from the origin along the y-axis.

Figure 1.2 shows two number lines that are perpendicular to each other. The point at which they cross is called the *origin*. These lines and the plane they define are called the *rectangular coordinate axis system*.

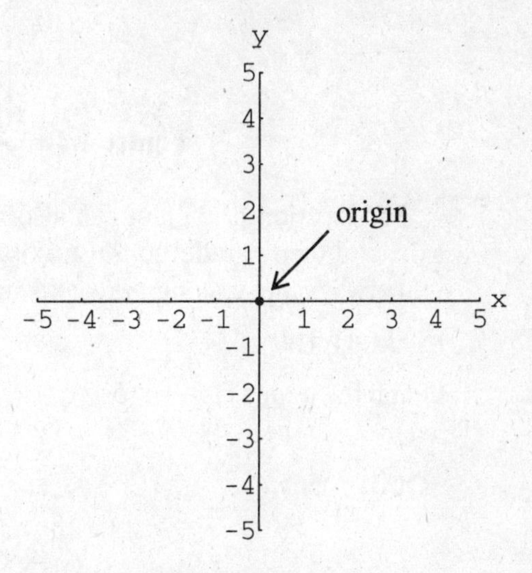

Figure 1.2 The Rectangular Coordinate System

The location of a point is defined as movement from the origin left or right and up or down. The x and y lines are called *axes*. The x-axis conveys horizontal movement and the y-axis conveys vertical movement.

The point defined by the ordered pair (3, 4) is located 3 units to the *right* of the origin and 4 units "*above*" the x-axis. The point (–1, –2) is one unit to the *left* of the origin and two units "*below*" the x-axis.

Figure 1.3 shows both of these points.

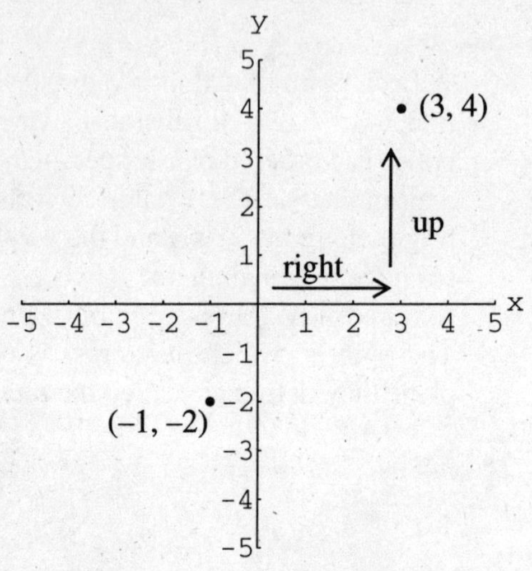

Figure 1.3 Graph of Points

The location of a point can also be found by moving first along the *y*-axis, and then parallel to the *x*-axis. However, to establish a pattern, we will always move along the *x*-axis first.

EXAMPLE 1.4

Graph the points (1, –3); (5, 2); (–4, 0) and $\left(-\frac{3}{2}, 4\right)$.

SOLUTION 1.4

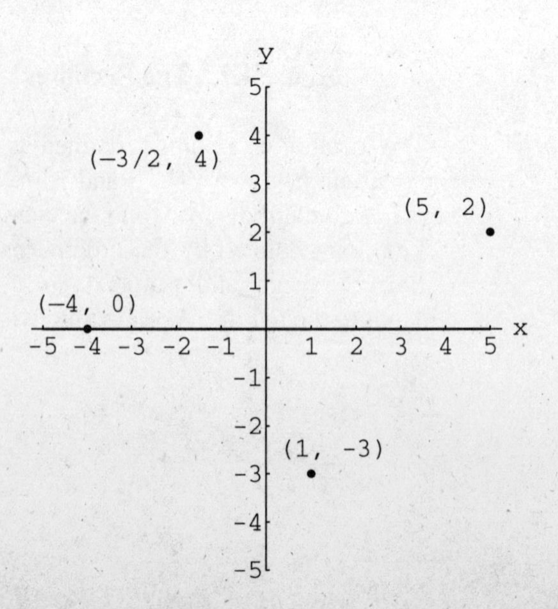

1.4 GRAPHING LINEAR FUNCTIONS

The equation of a line can be used to obtain points that are on the graph of the line and to find the slope of the line. This information can be used, in turn, to draw the graph of the line.

Three methods of graphing will be discussed:

(a) Using a table (three points).
(b) Using the slope and y-intercept.
(c) Using the x-intercept and y-intercept.

Graphing from a Table

There are essentially four parts in the process in using a table. Given an equation of a line, we:

(1) solve for y in terms of x (if needed).
(2) construct a table of three points by substituting numbers for x.
(3) plot the points.
(4) draw a line connecting the points.

EXAMPLE 1.5

Use a table to graph the line for the linear funtions shown.

(a) $2x + y = 3$
(b) $x - 3y = 6$

SOLUTION 1.5

(a) Given the equation $2x + y = 3$.

(1)
$$\begin{array}{rcl} 2x + y &=& 3 \\ \underline{-2x } && \underline{-2x} \\ y &=& -2x + 3 \end{array}$$

Solve for y in terms of x. Place y on one side of the equal and x and all other numbers on the other side.

(2) Set up a table by picking **small** integers for x.

x	y
-1	
0	
1	

$y = -2x + 3$

$y = -2(0) + 3 = 0 + 3 = 3$
$y = -2(1) + 3 = -2 + 3 = 1$
$y = -2(-1) + 3 = 2 + 3 = 5$

Substitute the three x-values in the equation from part 1. Usually a value of 0 for x and other small whole numbers are adequate.

Complete the table.

x	y
0	3
1	1
–1	5

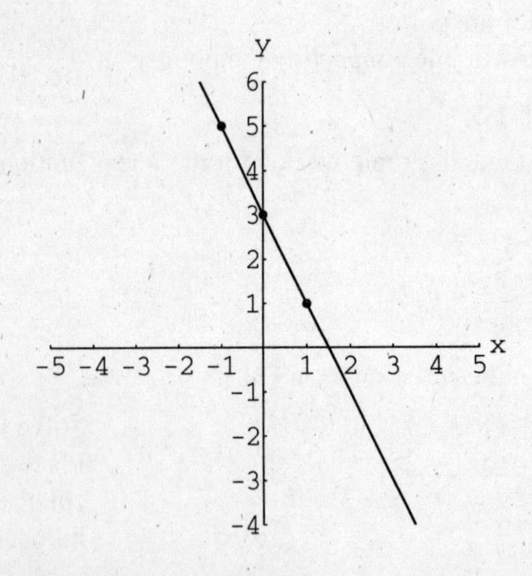

(3) Plot the points.
(4) Connect the points with a line. The third point serves as a check.

(b) $x - 3y = 6$

Given equation.

(1) $x - 3y = 6$

$$\begin{array}{r} x - 3y = 6 \\ \underline{-x \qquad\quad -x} \\ -3y = 6 - x \end{array}$$

$$\frac{-3y}{-3} = \frac{6}{-3} - \frac{x}{-3}$$

$$y = \frac{1}{3}x - 2$$

Solve the equation for y in terms of x.

(2) $y = \frac{1}{3}x - 2$

Substitute three values of x into the equation to build a table. Since the denominator of the slope is 3, use multiples of 3.

x	y
0	
3	
6	

$$y = \frac{1}{3}(0) - 2 = -2$$
$$y = \frac{1}{3}(3) - 2 = 1 - 2 = -1$$
$$y = \frac{1}{3}(6) - 2 = 2 - 2 = 0$$

Complete the table.

x	y
0	–2
3	–1
6	0

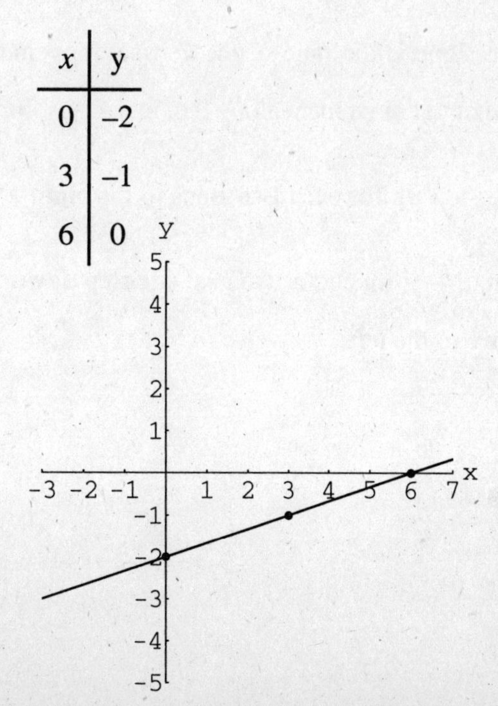

(3) Plot the points.
(4) Graph the line.

Using the Slope and the y-intercept

Another method of graphing a line is to start at the y-intercept and use the slope to find the next point. This procedure is most easily used when the equation to be graphed is in the **slope-intercept** form.

EXAMPLE 1.6

Use the slope and the y-intercept to graph:

(a) $y = \frac{1}{2}x + 3$

(b) $y = -2x - 1$

(c) $y = -\frac{3}{4}x + 1$

Note: These equations have already been placed in slope-intercept form. We discussed this process earlier in the chapter. We must solve for y in terms of x.

SOLUTION 1.6

(a) The equation, $y = \frac{1}{2}x + 3$, is in the slope-intercept form. The slope (m) is $\frac{1}{2}$ and the y-intercept (b) is 3.

The slope is the ratio of the vertical movement along the line to the horizontal movement along the line. Thus, for a slope of $\frac{1}{2}$, we move **one** step **up** for each **two** steps to the **right**. Since $\frac{1}{2}$ is also equivalent to $\frac{-1}{-2}$, we could also take **one** step **down** for each **two** steps if we move to the **left**.

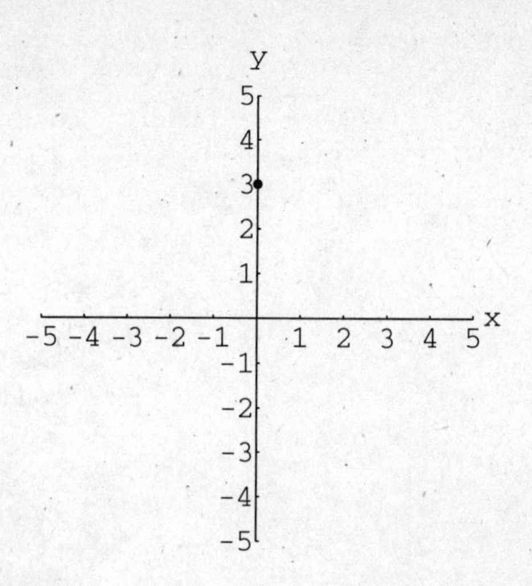

Set up grid and locate the *y*-intercept (0, 3).

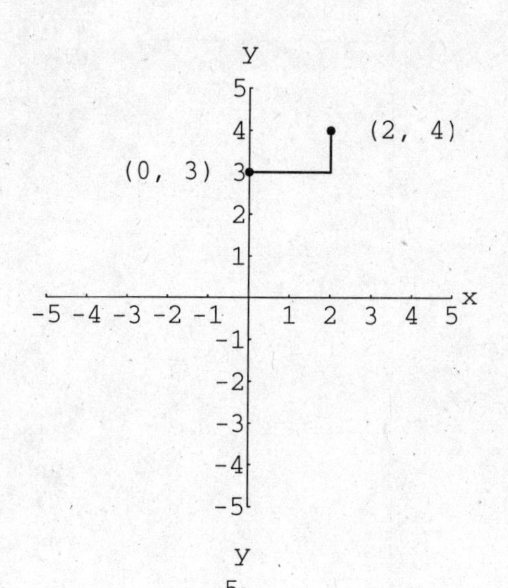

Use change in *x* and *y* determined by the slope and move 2 steps right and 1 step up.

Place second dot at $(0 + 2, 3 + 1)$ or $(2, 4)$.

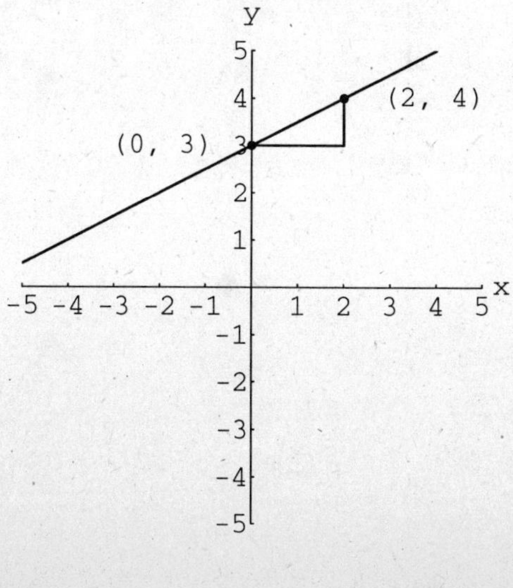

Draw the line $y = \frac{1}{2}x + 3$.

(b) The equation $y = -2x - 1$ has a y-intercept of -1 and a slope of -2 or $\frac{-2}{1}$ or $\frac{2}{-1}$. We start at -1 on the y axis and move two steps **down** and one step **right**. We could also move two steps **up** and one step **left**.

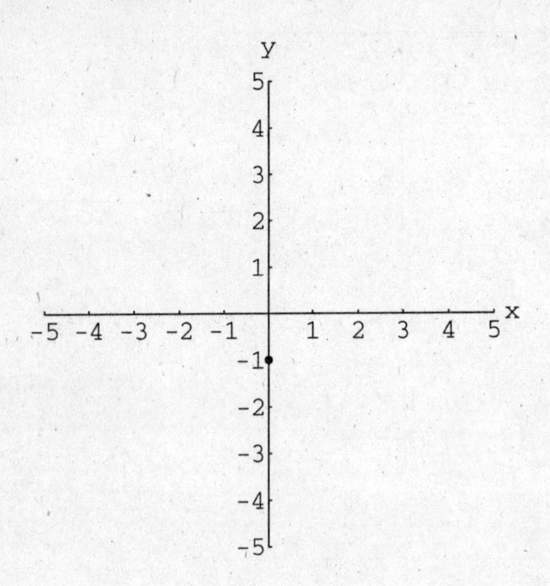

Start at the y-intercept of -1 on the y-axis.

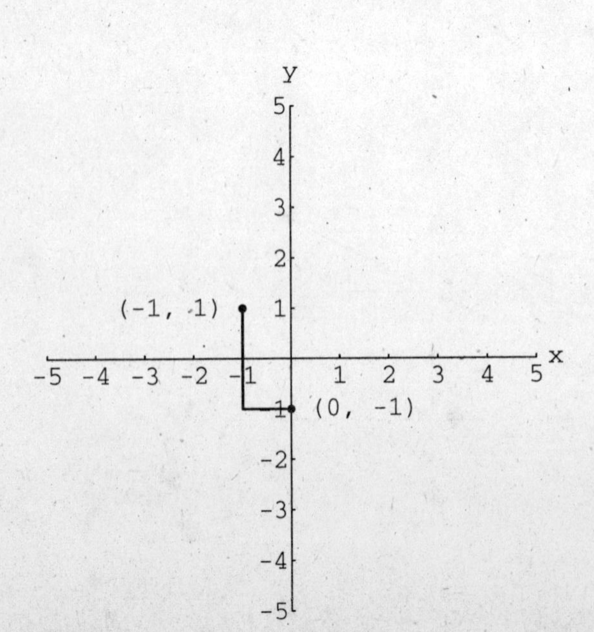

Move 1 unit left (-1) and 2 units up. This gives us $(0 - 1, -1 + 2) = (-1, 1)$ for the second point.

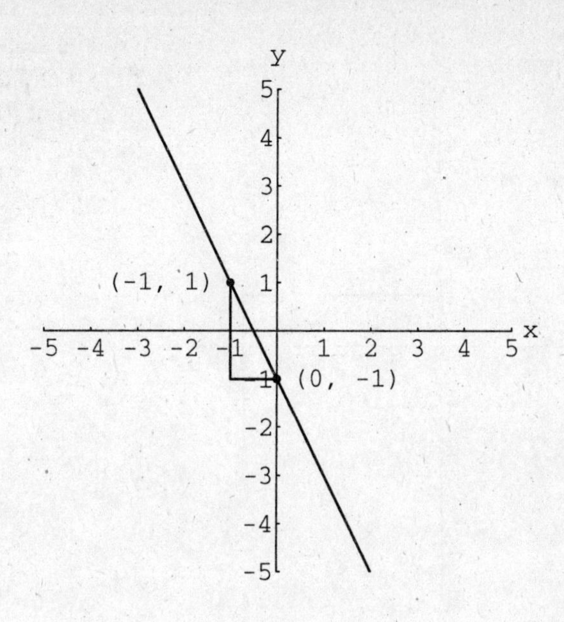

Draw the line connecting the two points. This is the line $y = -2x - 1$.

(c) The equation is $y = -\frac{3}{4}x + 1$. Plotting the line starts at 1 and moves 3 steps **down** for each 4 steps **right** (or 3 **up** and 4 **left**).

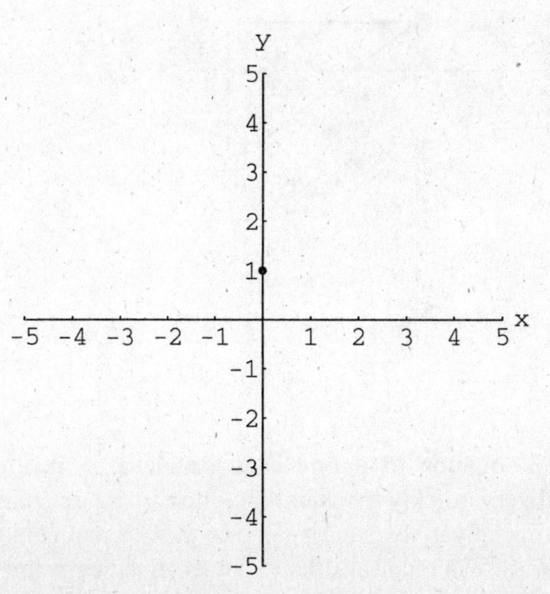

Start graphing at the y-intercept of 1.

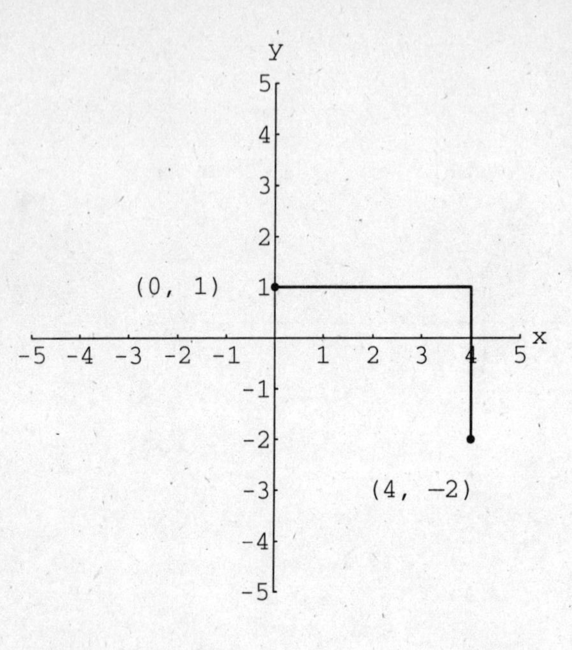

Use the slope and move right 4 and down 3 to the next point $(0 + 4, 1 - 3) = (4, -2)$.

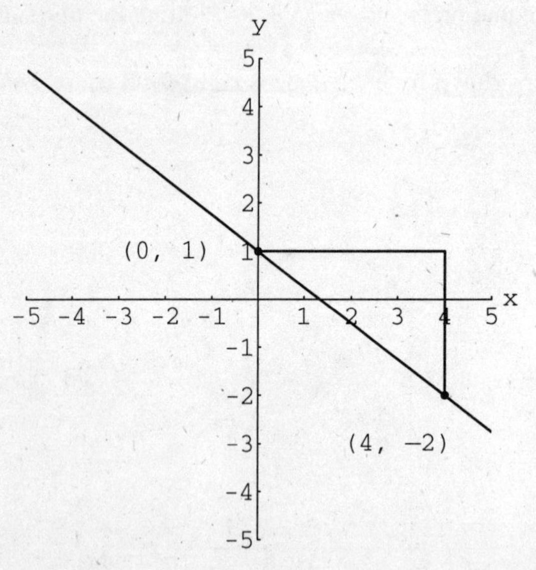

Draw the line

$$y = \left(-\frac{3}{4}\right)x + 1$$

Using the x-intercept and the y-intercept

If the equation of a line is in standard form, the equation can be graphed very quickly by examining the axis intercepts. In this process we set one of the variables (x or y) equal to zero and solve for the other variable in turn. For example, the variables in the equation $4x + 3y = 8$, can be set in turn to zero to yield:

$$4(0) + 3y = 8$$
$$0 + 3y = 8$$
$$3y = 8$$ <-- notice that the x-term is gone.
$$y = \frac{8}{3}$$

When we find that $y = \frac{8}{3}$, we are locating a point on the y-axis that is known as the y-intercept of the line. Similarly, if we set the y-value equal to zero, we find the x-intercept.

$$4x + 3(0) = 8$$
$$4x + 0 = 8$$
$$4x = 8$$ <-- notice that the y-term is gone.
$$x = 2$$

If you visually eliminate the other term by putting your finger over it as you solve for the x- or y-intercept, you are duplicating the process of setting the term to zero. This, perhaps, makes the process slightly faster.

EXAMPLE 1.7

Use the x-intercept and the y-intercept of the lines below to aid in graphing the lines.

(a) $3x + 2y = 6$
(b) $x - 2y = 4$
(c) $2x - 5y = 4$

SOLUTION 1.7

With the equation already in standard form, the best course for graphing is to quickly find the intercepts.

(a) The equation for this problem is $3x + 2y = 6$.

$3x + 2y = 6$	Write the given equation.
$3(0) + 2y = 6$	Replace x with 0 and solve for y.
$0 + 2y = 6$ $2y = 6$ $y = \frac{6}{2} = 3$	We find one intercept and set it aside and solve for the other intercept.

$$3x + 2(0) = 6$$
$$3x + 0 = 6$$
$$3x = 6$$
$$x = \frac{6}{3} = 2$$

Similarly, solve for the other intercept.

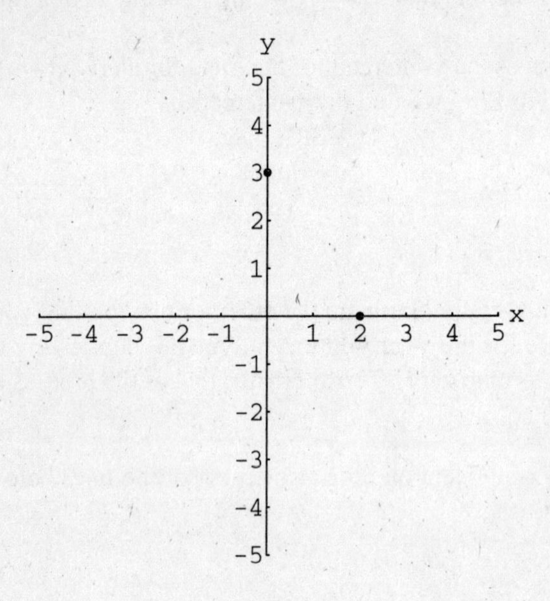

Plot both intercept points (0, 3) and (2, 0).

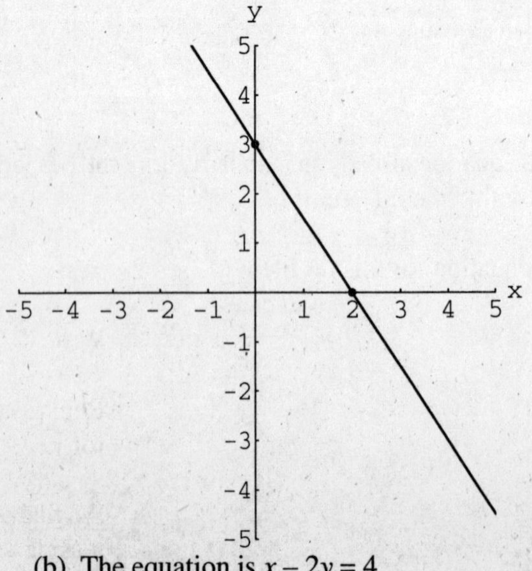

Draw the line through the points.

(b) The equation is $x - 2y = 4$.

$$x - 2y = 4$$

Write the given equation.

$$0 - 2y = 4$$

Replace x with 0 and solve for y.

$$-2y = 4$$

$$y = \frac{4}{-2} = -2$$

We find one intercept and set it aside and solve for the other intercept.

$$x - 2(0) = 4$$
$$x - 0 = 4$$
$$x = 4$$

Similarly, solve for the other intercept.

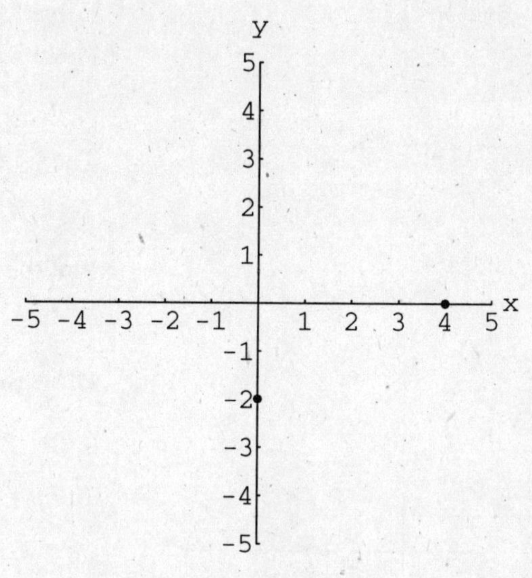

Plot the points (0, – 2) and (4, 0).

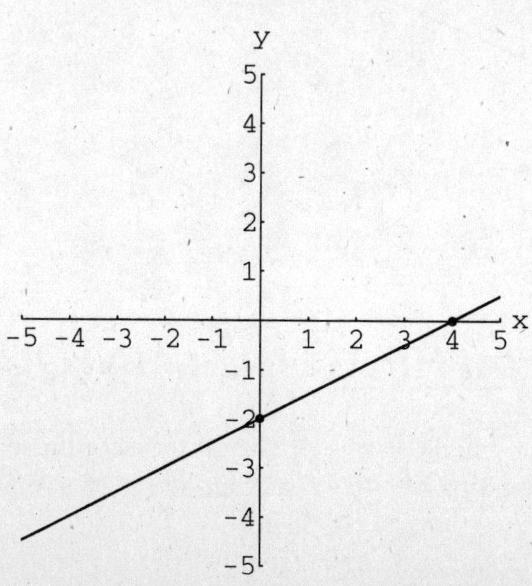

Draw the line through the two points.

(c) The equation to be graphed is $2x - 5y = 4$.

$2x - 5y = 4$	Write the given equation.
$2x - 5(0) = 4$ $2x - 0 = 4$	Replace y with 0 and solve for x.
$2x = 4$	Divide by two.
$x = \frac{4}{2} = 2$	The x-intercept is 2.
$2(0) - 5y = 4$ $0 - 5y = 4$ $-5y = 4$	Solve for the y-intercept by placing a 0 in for x.
$y = \frac{-4}{5}$	The y-intercept is $\frac{-4}{5}$.

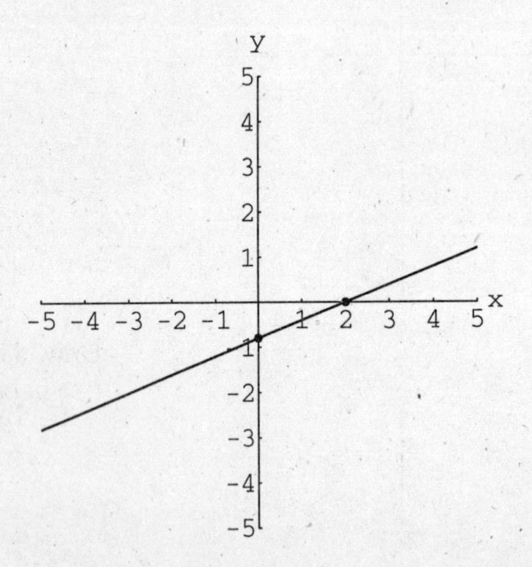

Plot the two intercepts 2 and $-4/5$.

Draw the line $2x - 5y = 4$.

1.5 GRAPHING LINEAR INEQUALITIES

When a straight line is drawn on the coordinate axis plane, it divides the plane into two parts. The solution set of a linear inequality includes one of the two parts of the plane.

Linear inequalities are inequalities of the forms:

$$ax + by \geq c$$

$$ax + by \leq c$$

$$ax + by > c$$

$$ax + by < c$$

To display the inequality on a graph, we first draw the line that would result if the inequality were an equation. The region of the graph on one side of the line that makes the inequality true is then shaded in some manner to highlight it.

EXAMPLE 1.8

Graph the following linear inequalities.

(a) $3x + 2y < 6$
(b) $x - y > 1$
(c) $4x - 2y \leq 8$
(d) $3x + y \geq 2$

SOLUTION 1.8

The solution process for each of these linear inequalities involves the graphing of the related linear equation. We use a **dotted line** for those linear inequalities with a $<$ or $>$ symbol to show that the points on the line are not part of the solution. We use a **solid line** for those linear inequalities with a $\leq$ or $\geq$ symbol to include those points in the solution.

The two areas are tested using one point on the plane, and the proper area is shaded.

(a) The inequality is $3x + 2y < 6$.

$3x + 2y < 6$		Write the given inequality.
$3x + 2y = 6$		Rewrite the inequality as a linear **equation**.
$2y = 6$	$3x = 6$	
$y = 3$	$x = 2$	Solve for the x- and y-intercepts.

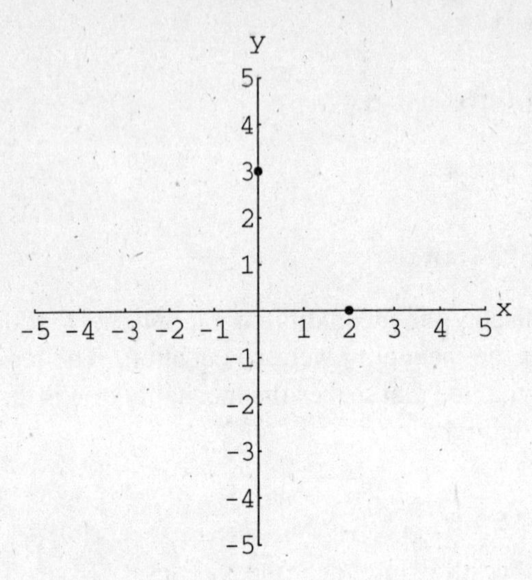

Plot the points at x-intercept (2, 0) and y-intercept (0, 3).

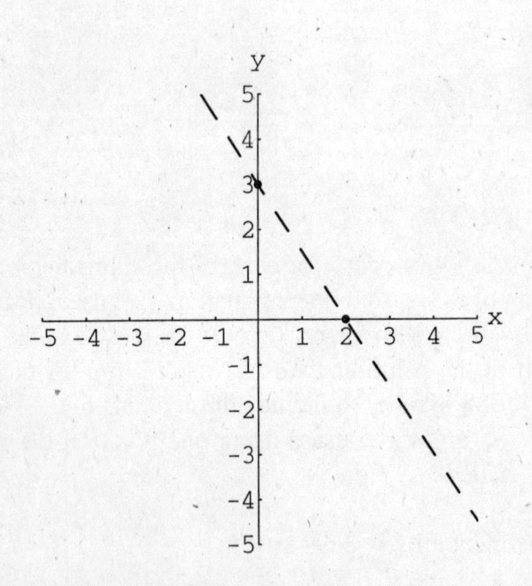

Since the inequality symbol is a <, we plot a dotted line.

$$3x + 2y < 6$$
$$3(0) + 2(0) < 6 \quad ?$$
$$0 < 6$$
TRUE

Test the inequality by putting the coordinates of the origin into the inequality. The solution is checked to see if it yields a true or false statement.

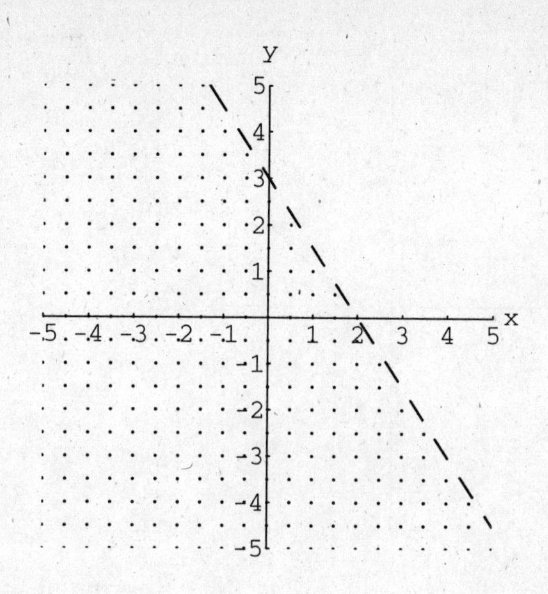

The area that contains the origin gave us a TRUE inequality, so we shade the area with the origin.

(b) The given inequality is $x - y > 1$.

$x - y > 1$

Write the given inequality.

$x - y = 1$

Rewrite the inequality as an **equation**.

$x = 1 \qquad -y = 1$

$\qquad\qquad\; y = -1$

Solve for the intercepts.

Plot the intercepts (1, 0) for the x-intercept and (0, –1) for the y-intercept.

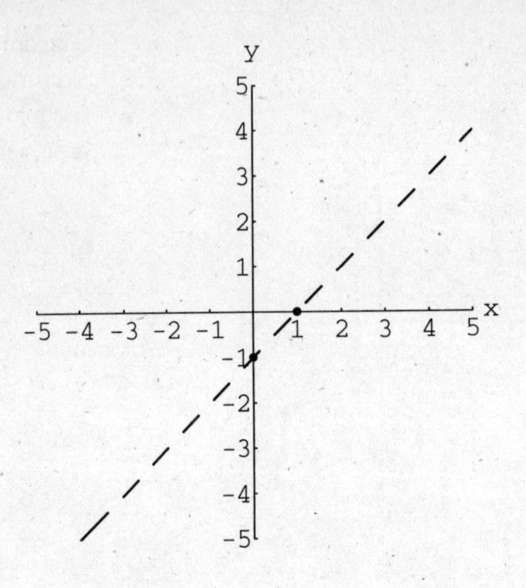

Draw a dotted line because of the > sign.

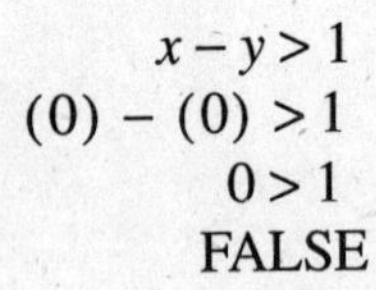

$$x - y > 1$$
$$(0) - (0) > 1$$
$$0 > 1$$
FALSE

Test the inequality by putting the coordinates of the origin into the inequality. The solution is checked to see if it yields a true or false statement.

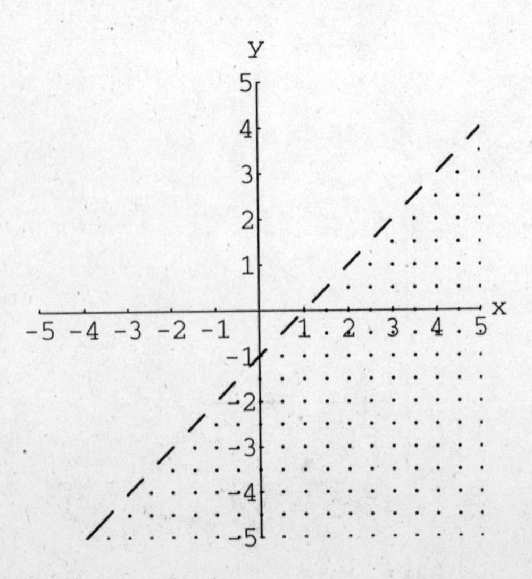

Since the test is FALSE, the area which does not contain the origin is shaded.

(c) The inequality for this problem is $4x - 2y \le 8$.

$4x - 2y \le 8$ — Copy the given inequality.

$4x - 2y \le 8$ — Rewrite the inequality as an equation.

$4x = 8$ $-2y = 8$

$x = 2$ $y = -4$ — Solve for the intercepts.

Plot the intercepts (2, 0) and (0, –4).

Draw a solid line since the inequality symbol is $\le$.

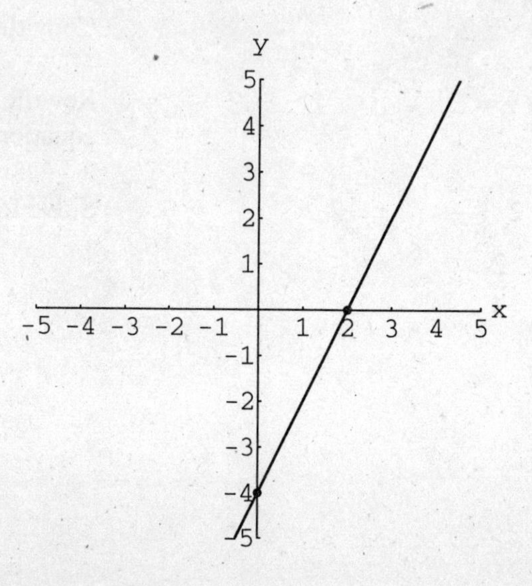

$$4x - 2y \leq 8$$
$$4(0) - 2(0) \leq 8 \quad ?$$
$$0 \leq 8$$
TRUE

Test the inequality by putting the coordinates of the origin into the inequality. The solution is checked to see if it yields a true or false statement.

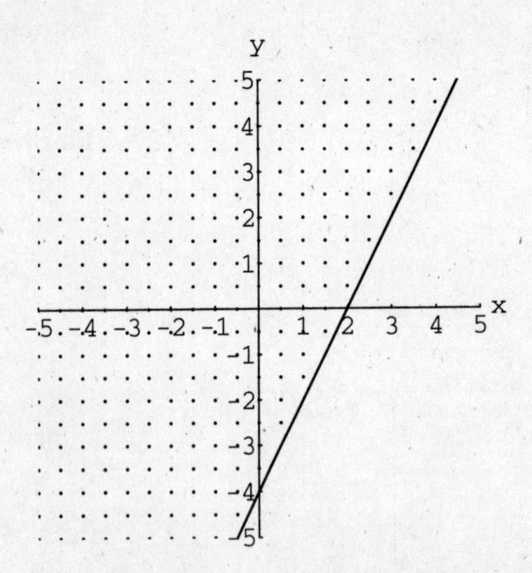

The test is TRUE, thus we shade the area WITH the origin.

(d) The inequality for this problem is $3x + y \geq 2$.

$3x + y \geq 2$

Copy the inequality.

$3x + y = 2$

Rewrite the inequality as an equation.

$3x = 2 \qquad y = 2$

$x = \dfrac{2}{3}$

Solve for the intercepts.

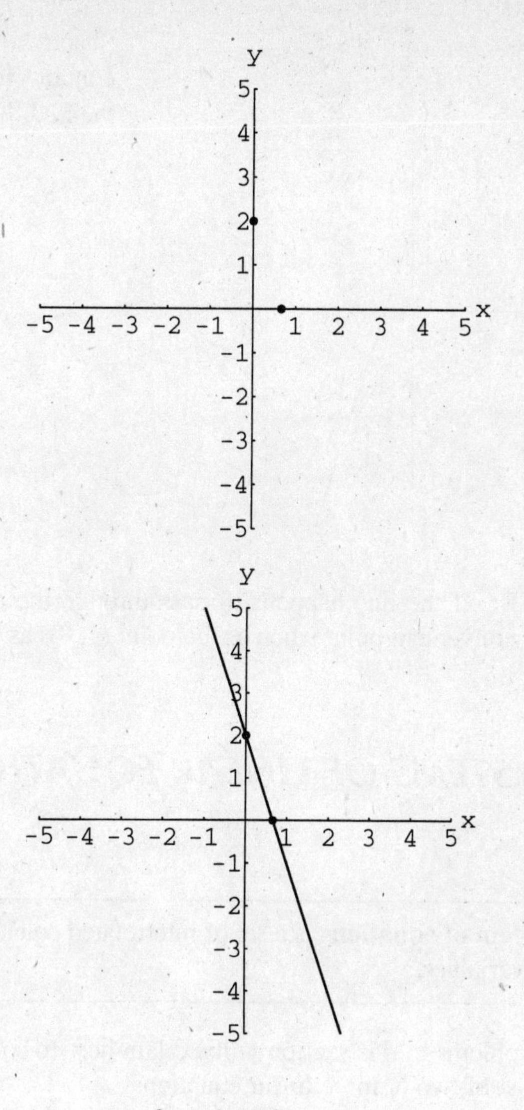

Plot the intercepts of (0, 2) for the *y*-intercept and (2/3, 0) for the *x*-intercept.

Draw a solid line since the inequality symbol is $\geq$.

$$3x + 1y \geq 2$$

$$3(0) + 1(0) \geq 2 \quad ?$$

$$0 \geq 8$$

FALSE

Test the inequality by putting the coordinates of the origin into the inequality. The solution is checked to see if it yields a true or false statement.

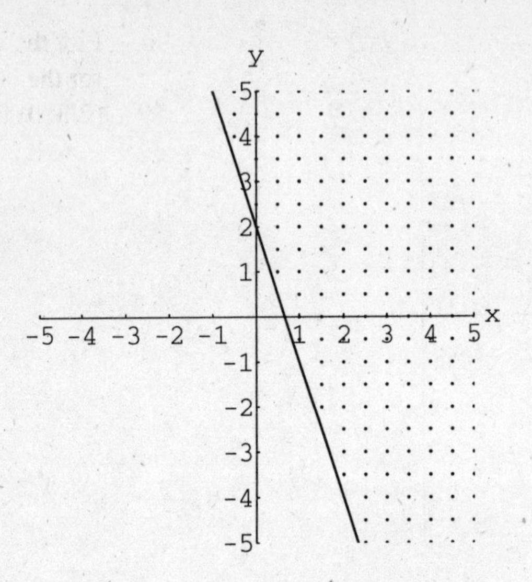

Shade the area that does not contain the origin because of the FALSE test.

NOTE: If the line happens to pass through the origin, select some other convenient point (such as the point (1, 0)) as the test point.

1.6 SYSTEMS OF LINEAR EQUATIONS

> A **system of equations** is a set of interrelated equations with common variables.

The problems in this section will explain how to solve some problems that consist of two or more linear equations.

Systems of Two Equations

An example of a system with two equations is the following:

$$\begin{aligned} 3x + 2y &= 7 \\ x - 3y &= -5 \end{aligned}$$

Note that the equations are written one above the other. This is the common format for such systems, and, as we shall see later, helps in the solution process.

Each of the two equations in the system can be represented on a two-dimensional graph as a straight line. Lines in two dimensions must relate to each other in one of three ways:

(a) they intersect each other at one unique point.

(b) they are different forms of the same line.

(c) they run parallel to each other.

Consistent System

The solution of the system above is the point (1, 2) on the coordinate axis system. This type of system is considered to be **consistent**, in that it only has one solution. For a system of two equations with two variables, the solution is an ordered pair (x, y). This solution represents the intersection of two lines. Figure 1.4 shows the graph of a consistent system.

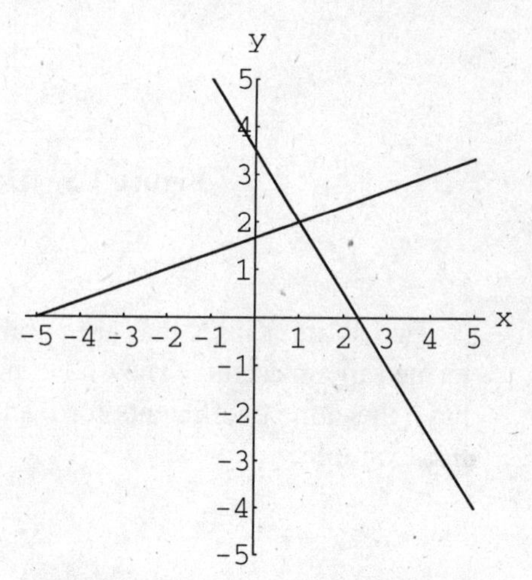

Figure 1.4 Consistent System

Dependent System

When two linear equations describe the same line, they are called **dependent**. The equations initially may appear to be equations of different lines, but are actually multiples of one another.

(A) $3x + 2y = 6$
(B) $15x + 10y = 30$

The above equations are the same with equation B five times equation

A. They both reduce to $y = -\frac{3}{2}x + 3$. Since these two equations are the same line, they are considered to have an infinite number of intersection points.

Figure 1.5 shows an example of a dependent system.

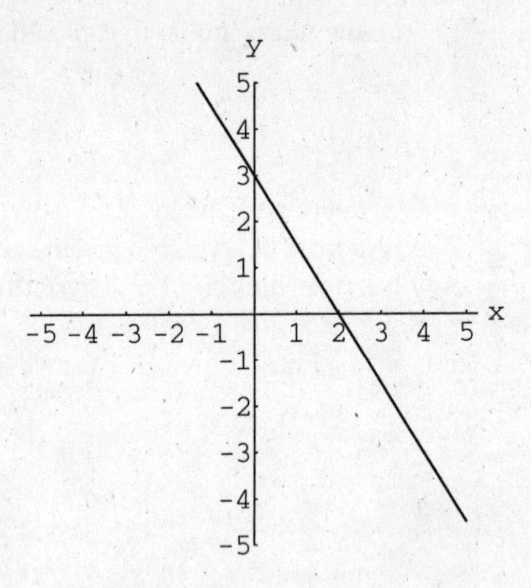

Figure 1.5 Dependent System

Inconsistent System

Two straight lines that are parallel to each other form a system that is termed inconsistent. They have no points of intersection. Parallel lines have the same coefficients for x and y, but have different constant values. For example,

$$5x - 4y = 8$$
$$\text{and} \quad 5x - 4y = -8$$

are parallel lines. Figure 1.6 shows an inconsistent system.

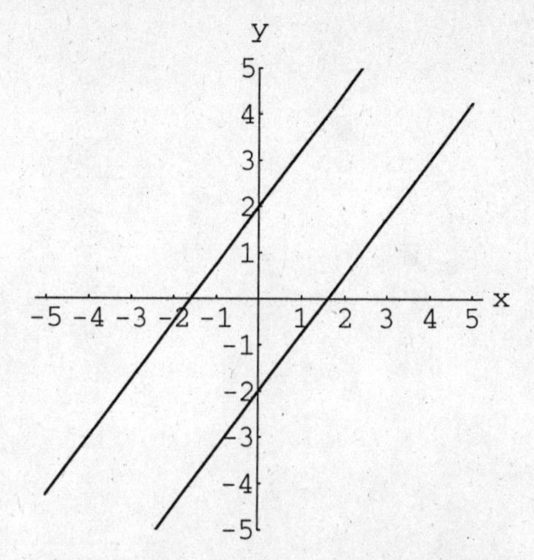

Figure 1.6 Inconsistent System

Solution Methods

The two usual methods for solving systems of two equations are (a) the **substitution** method and (b) the **addition (elimination)** method. Both of these processes are demonstrated in the following examples.

Substitution Method

The substitution method for systems of equations begins by solving one of the equations for one of the variables in terms of the other. The solution replaces the solved-for variable in the second equation. With the second equation now containing only one variable, the value of that variable can be found. Either equation can then be used to find the value of the second variable.

The substitution method is shown in the next example.

EXAMPLE 1.9

Solve the following system of equations using the *substitution method.*

(a) $3x - y = 7$
$x + y = 5$

(b) $x - 2y = 1$
$2x + y = 5$

(c) $5x + 7y = 13$
$x + 6y = 8$

(d) $4x - 2y = 3$
$8x - 4y = 6$

SOLUTION 1.9

(a)

(i) $3x - y = 7$ (ii) $x + y = 5$	Copy both equations.
$x + y = 5$ $y = 5 - x$	Solve for y by subtracting x from both sides of the equation (ii).
$3x - y = 7$ $3x - \mathbf{(5 - x)} = 7$	Substitute for y in the **other** equation (i).
$3x - 5 + x = 7$	Make sure to distribute -1 through $5 - x$.
$4x - 5 = 7$	Add 5 to both sides of equation.
$4x = 12$	Divide by 4.
$x = 3$	We now know that $x = 3$.
$x + y = 5$ $3 + y = 5$ $y = 2$	Once the value of x is known, y can be determined by substituting the value of x in either equation.

The solution set is **(3, 2)**.

(b)

(i) $x - 2y = 1$ (ii) $2x + y = 5$	Copy both equations.
$2x + y = 5$ $y = 5 - 2x$	Solve for y by subtracting $2x$ from both sides of the equation (ii).
$x - 2y = 1$ $x - 2\mathbf{(5 - 2x)} = 1$	Substitute for y in the **other** equation (i).

$x - 10 + 4x = 1$	Make sure to distribute –1 through $5 - 2x$.
$5x - 10 = 1$	Add 10 to both sides of equation.
$5x = 11$	Divide by 5.
$x = \frac{11}{5}$	We now know that $x = 11/5$.
$2x + y = 5$ $(2)\frac{11}{5} + y = 5$	Once the value of x is known, y can be determined by substituting the value of x in an equation.
$\frac{22}{5} + y = 5$	
$\frac{22}{5} + y = \frac{25}{5}$	
$y = \frac{25}{5} - \frac{22}{5} = \mathbf{\frac{3}{5}}$	

The solution set is $\left(\mathbf{\frac{11}{5}, \frac{3}{5}}\right)$.

(c)

(i) $5x + 7y = 13$ (ii) $x + 6y = 8$	Copy both equations.
$x + 6y = 8$ $x = 8 - 6y$	Solve for x by subtracting $6y$ from both sides of the equation (ii).
$5x + 7y = 13$ $5(\mathbf{8 - 6y}) + 7y = 13$	Substitute for x in the **other** equation (i).
$40 - 30y + 7y = 13$	Distribute 5 through $8 - 6y$.
$40 - 23y = 13$	Subtract 40 on both sides of equation.

$-23y = -27$	Divide by –23.
$y = \dfrac{-27}{-23}$	Now the variable, *y*, is known.
$\mathbf{y = \dfrac{27}{23}}$	
$x + 6y = 8$ $x + (6)\cdot\dfrac{27}{23} = 8$	Once the value of *y* is known, *x* can be determined by substituting the value of *y* in an equation.
$x + \dfrac{162}{23} = 8$	Multiply.
$\dfrac{162}{23} + x = \dfrac{184}{23}$	Find a common denominator Convert 8 to 184/23.
$x = \dfrac{184}{23} - \dfrac{162}{23} = \mathbf{\dfrac{22}{23}}$	Subtract 162/23 from both sides.

The solution set is $\left(\mathbf{\dfrac{22}{23}, \dfrac{27}{23}}\right)$.

(d)

(i) $4x - 2y = 3$ (ii) $8x - 4y = 6$	Copy both equations.
$8x - 4y = 6$ $8x = 6 + 4y$ $x = \dfrac{3}{4} + \dfrac{1}{2}y$	Solve for *x* in terms of *y* by adding 4*y* and dividing by 8 in equation (ii).
$4x - 2y = 3$ $4\left(\mathbf{\dfrac{3}{4}} + \mathbf{\dfrac{1}{2}}y\right) - 2y = 3$	Substitute for *x* in the **other** equation (i). Distribute 4 through (3/4 + *y*/2).
$3 + 2y - 2y = 3$	Combine like terms.

$3 + 0 = 3$ Simplify.

$3 = 3$ The answer is an identity.

This type of answer means that there are an infinite number of solutions. The solutions to a dependent system will end in the form of $a = a$, where a is a constant.

Addition Method

There are three operations that are allowed within systems of equations and they will be used in the **addition (or elimination)** method:

(1) The equations can be **arranged** in any order top to bottom.

(2) The numerical coefficients can be **adjusted** by any multiple of the equation for a given equation without changing the values of the variables in that equation.

(3) Two equations can be **added** together to form a new equation to create an equivalent system.

EXAMPLE 1.10

Solve the following system of equations using the **addition (elimination) method.**

(a) $x - y = 6$
$x + y = 14$

(b) $3x + 2y = 7$
$x + 2y = 3$

(c) $3x - 5y = 6$
$2x + 2y = 13$

(d) $5x - 3y = 11$
$10x - 6y = 3$

SOLUTION 1.10

(a)

(i) $x - y = 6$ Copy the given equations.
(ii) $x + y = 14$

$$\begin{array}{r} x - y = 6 \\ \underline{x + y = 14} \\ 2x + 0 = 20 \end{array}$$

Add the two equations vertically. $x + x = 2x$, $-y + y = 0$, $6 + 14 = 20$.

$$\frac{2x}{2} = \frac{20}{2}$$

Divide by 2.

$$\mathbf{x = 10}$$

The value of x is known.

$$\begin{array}{r} x - y = 6 \\ 10 - y = 6 \end{array}$$

Substitute 10 for x in equation (i).

$$-y = 6 - 10 = -4$$

Subtract 10 from both sides of equation (i).

$$\mathbf{y = 4}$$

The solution set is **(10, 4)**.

(b)

(i) $3x + 2y = 7$
(ii) $x + 2y = 3$

Copy the given equations.

$$\begin{array}{c} 3x + 2y = 7 \\ (-3)(x + 2y) = (-3)(3) \end{array}$$

Equation (ii) is adjusted by multiplying by -3 so that the coefficients of x are the same magnitude but opposite in sign.

$$\begin{array}{r} 3x + 2y = 7 \\ \underline{-3x - 6y = -9} \\ 0 + -4y = -2 \end{array}$$

Add the previous two equations vertically. $3x + (-3x) = 0$, $2y - 6y = -4y$, $7 - 9 = -2$.

$$\frac{-4y}{4} = \frac{-2}{-4}$$

Divide by -4.

$$\mathbf{y = 1/2}$$

The variable, y, is known.

$$x + 2\left(\frac{1}{2}\right) = 3$$

Substitute for y in equation (ii).

$$\begin{array}{r} x + 1 = 3 \\ \mathbf{x = 2} \end{array}$$

Subtract 1 from both sides of equation (ii).

The solution set is $\left(\mathbf{2}, \mathbf{\frac{1}{2}}\right)$.

(c)

(i) $3x - 5y = 6$ (ii) $2x + 2y = 13$	Copy the given equations.
$3x - 5y = 6$ $2x + 2y = 13$	In this case, both equations must be **adjusted** so that one variable can be eliminated. The y will be eliminated here.
$(2)(3x - 5y) = (2)(6)$ $(5)(2x + 2y) = (5)(13)$	Multiply equation (i) by 2 and equation (ii) by 5.
$6x - 10y = 12$ $\underline{10x + 10y = 65}$ $16x + 0 \quad = 77$	Add vertically so that the y variable is eliminated. $6x + 10x = 16x$, $-10y + 10y = 0$, $12 + 65 = 77$.
$16x = 77$ $\frac{16x}{16} = \frac{77}{16}$	Divide by 16 to find x.
$\boldsymbol{x = \frac{77}{16}}$	Now x is known.
$2\left(\boldsymbol{\frac{77}{16}}\right) + 2y = 13$	Substitute the x-value in equation (ii).
$\frac{154}{16} + 2y = 13 = \frac{208}{16}$	Find a common denominator Convert 13 to 208/16.
$2y = \frac{208}{16} - \frac{154}{16} = \frac{54}{16}$	Subtract 154/16 from both sides of equation.
$2y = \frac{54}{16}$	
$\boldsymbol{y} = \frac{54}{32} = \boldsymbol{\frac{27}{16}}$	Solve for y by dividing by 2.

The solution set is $\left(\boldsymbol{\frac{77}{16}, \frac{27}{16}}\right)$.

(d)

(i) $5x - 3y = 11$ (ii) $10x - 6y = 3$	Copy the given equation.
$(-2)(5x - 3y) = (-2)(11)$ $10x - 6y = 3$	Multiply equation (i) by –2 so that the coefficients of x add to 0.
$-10x + 6y = -22$ $\underline{10x - 6y = 3}$ $0 + 0 = -19$	Add the two equations together. $-10x + 10x = 0$, $6y - 6y = 0$, $-22 + 3 = -19$. Note that the "solution" is the nonsense equation $0 = -19$.

There is **no solution** to this system.
This is an **inconsistent system**.

System of Three Equations

Two ways to solve a system of two equations were presented in the previous pages. Now we will focus on a method to solve a system of three equations. It should be emphasized that this is not the only way to solve such systems, but, perhaps, is one of the more straightforward.

The process we will use is an extension of the **addition** method. Recall that the equations can be **arranged** in any order from top to bottom, and that two equations may be **added** to form a third equation. This new equation can replace one of the other two. Finally, any of the equations can be multiplied by a factor to **adjust** the size of the coefficients.

EXAMPLE 1.11

Find the solution set for the following systems of three equations.

(a)
$$\begin{aligned} 3x - y - z &= 2 \\ x + y + z &= 2 \\ 2x - y + 2z &= -2 \end{aligned}$$

(b)
$$\begin{aligned} x + 4y - z &= 5 \\ 3x - y - 2z &= -3 \\ x + 5y + z &= -4 \end{aligned}$$

(c)
$$\begin{aligned} x - y + z &= 2 \\ 2x - y + 4z &= 2 \\ x + y + z &= 0 \end{aligned}$$

SOLUTION 1.11

(a)

(i) $3x - y - z = 2$ (ii) $x + y + z = 2$ (iii) $2x - y + 2z = -2$	Copy the given equations.
(i) $3x - y - z = 2$ (ii) $\underline{x + y + z = 2}$ $4x + 0 + 0 = 4$	Add equations (i) and (ii) to find a sum with both y and z eliminated.
$4x = 4$ $\mathbf{x = 1}$	Divide by 4. We find the solution to x.
(ii) $x + y + z = 2$ (iii) $\underline{2x - y + 2z = -2}$ $3x + 0 + 3z = 0$	Equation (ii) is added to equation (iii) to form a new equation without y.
$3x + 3z = 0$ $3(\mathbf{1}) + 3z = 0$ $3 + 3z = 0$ $3z = -3$ $\mathbf{z = -1}$	The value for x found from the addition of equations (i) and (ii) is substituted into the equation for x and z.
$x + y + z = 2$ $\mathbf{1} + y + (\mathbf{-1}) = 2$ $y + 0 = 2$ $\mathbf{y = 2}$	The variables x and z are replaced by their numerical values to find the value for y.

The solution set is **(1, 2, –1)**.

(b)

(i) $x + 4y - z = 5$ (ii) $3x - y - 2z = -3$ (iii) $x + 5y + z = -4$	Copy the given equations.
$(-2)(x + 4y - z) = (-2)(5)$ $\underline{3x - y - 2z = -3}$	Multiply equation (i) by –2 and add it to equation (ii) to eliminate z.

$-2x - 8y + 2z = -10$ $\underline{3x - y - 2z = -3}$ (iv) $x - 9y + 0 = -13$	The top equation has been multiplied by –2 to eliminate z. The equations are added. A new equation, (iv), is formed.

(iv) $\mathbf{x - 9y = -13}$

$x + 4y - z = 5$ $\underline{x + 5y + z = -4}$ (v) $2x + 9y + 0 = 1$	Add equations (i) and (iii). This addition also eliminates z. This addition forms a new equation, (v).

(v) $\mathbf{2x + 9y = 1}$

(iv) $x - 9y = -13$ (v) $\underline{2x + 9y = 1}$ $3x + 0 = -12$	The next step is to add the two equations (iv) and (v). This addition eliminates y so that the value of x can be found.
$3x = -12$	Solve for x by dividing by 3.
$\mathbf{x = -4}$	
$x - 9y = -13$ $-4 - 9y = -13$	The value for x is substituted into equation (iv).
$-9y = -9$ $\mathbf{y = 1}$	Divide by –9.
$x + 5y + z = -4$ $\mathbf{-4} + 5(\mathbf{1}) + z = -4$	Select one of the three original equations to substitute x and y. In this case, we use equation (iii).
$-4 + 5 + z = -4$	Simplify.
$1 + z = -4$ $\mathbf{z = -5}$	Subtract 1 from both sides.

The solution set is **(–4, 1, –5)**.

(c)

(i) $x - y + z = 2$
(ii) $2x - y + 4z = 2$
(iii) $x + y + z = 0$

Copy the given equations.

$$\begin{array}{r} 2x - y + 4z = 2 \\ \underline{x + y + z = 0} \end{array}$$

(iv) $3x + 0 + 5z = 2$

Add the two equations, (ii) and (iii) to obtain an equation in x and z by eliminating y.

(iv) $\mathbf{3x + 5z = 2}$

(i) $x - y + z = 2$
(iii) $\underline{x + y + z = 0}$

(v) $2x + 0 + 2z = 2$

Equations (i) and (iii) are added to eliminate y.

(v) $\mathbf{2x + 2z = 2}$

$$\begin{array}{r} (-2)(3x + 5z) = (-2)(2) \\ \underline{(3)(2x + 2z) = (3)(2)} \end{array}$$

The equations must be adjusted to eliminate x. (iv) is multiplied by –2 and (v) is multiplied by 3.

$$\begin{array}{r} -6x - 10x = -4 \\ \underline{6x + 6z = 6} \\ 0 - 4z = 2 \end{array}$$

The value of z is found.

$$z = \frac{2}{-4} = \frac{\mathbf{-1}}{\mathbf{2}}$$

Divide by –4.

$$2x + 2z = 2$$

$$2x + 2\left(\frac{\mathbf{-1}}{\mathbf{2}}\right) = 2$$

The value for z is placed into equation (v).

$$\begin{array}{r} 2x + (-1) = 2 \\ 2x - 1 = 2 \\ 2x = 3 \end{array}$$

Simplify.
Add 1.

$$\mathbf{x = \frac{3}{2}}$$

Divide by 2.

$$x + y + z = 0$$

$$\left(\frac{3}{2}\right) + y + \left(\frac{-1}{2}\right) = 0$$

$$y + 1 = 0$$

$$\mathbf{y = -1}$$

The values for x and z are placed into one of the original equations, (iii), to find the value for y.

The solution set for this problem is $\left(\frac{3}{2}, -1, \frac{-1}{2}\right)$.

Practice Exercises

1. Find the slope of the line passing through the following points.

   (a) $(2, 3)$ and $(6, 1)$

   (b) $(2, 5)$ and $(-4, -1)$

   (c) $(8, -2)$ and $(1, -2)$

2. (a) Determine the equation of the line passing through the points $(4, 1)$ and $(-1, 4)$.

   (b) Find the equation of the line with slope of $1/4$ and a y-intercept of -3.

   (c) Write the equation of a line with x-intercept of 3 and passing through $(-2, 1)$.

   (d) Find the equation of a line passing through the origin and crossing $(3, -2)$.

   (e) Find the equation of a line with x-intercept of 7 and y-intercept of 4.

3. Solve the following inequalities:

   (a) $3x + 1 > 4$

   (b) $4x - 3 < 8$

4. Use a table to graph each line.

   (a) $2x + 4y = 8$

   (b) $x - y = 3$

5. Use the slope and the y-intercept to graph:

   (a) $y = \frac{1}{3}x + 2$

   (b) $y = -5x - 4$

6. Use the x-intercept and the y-intercept to graph each line.

   (a) $x + y = 2$

   (b) $x - 3y = 6$

   (c) $2x - 3y = 12$

7. Graph the following linear inequalities.

   (a) $x + 4y \leq 6$

   (b) $x - 2y > 3$

8. Solve each system of equations using the substitution method.

   (a) $x - y = 7$
   $x + y = 5$

   (b) $3x - 2y = 4$
   $2x + 2y = 6$

   (c) $3x + 3y = 6$
   $x + 6y = 8$

9. Solve each system of equations using the addition (elimination) method.

   (a) $x - y = 6$
   $x + y = 7$

   (b) $2x + 4y = 6$
   $x + 2y = 3$

   (c) $3x - 5y = 6$
   $x + y = 2$

10. Find the solution set for each system of three equations.

(a)
$$\begin{aligned} x - y - z &= 2 \\ x + y + z &= 12 \\ 2x - y - 5z &= 1 \end{aligned}$$

(b)
$$\begin{aligned} x + 4y - z &= 2 \\ 7x - y - z &= -3 \\ x + 5y + 3z &= 11 \end{aligned}$$

(c)
$$\begin{aligned} 4x - y + z &= 4 \\ 6x + 4z &= 14 \\ x + y + z &= 5 \end{aligned}$$

Answers

1. (a) –1/2
   (b) 1
   (c) 0

2. (a) $3x + 5y = 17$
   (b) $y = 1/4x - 3$
   (c) $x + 5y = 3$
   (d) $y = -2/3x$
   (e) $4x + 7y = 28$

3. (a) $x > 1$
   (b) $x < 5/4$

4. (a)

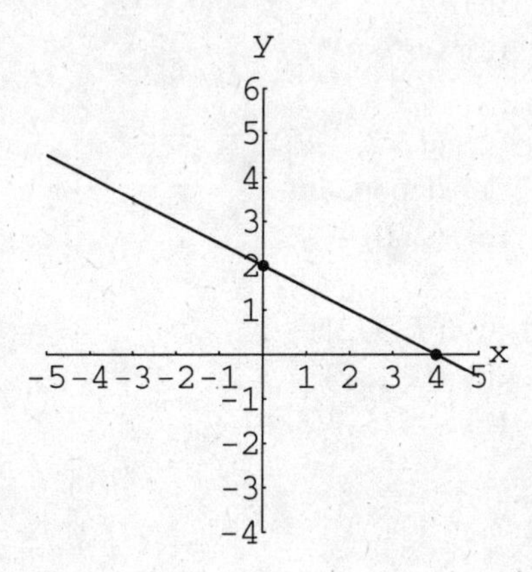

(b)

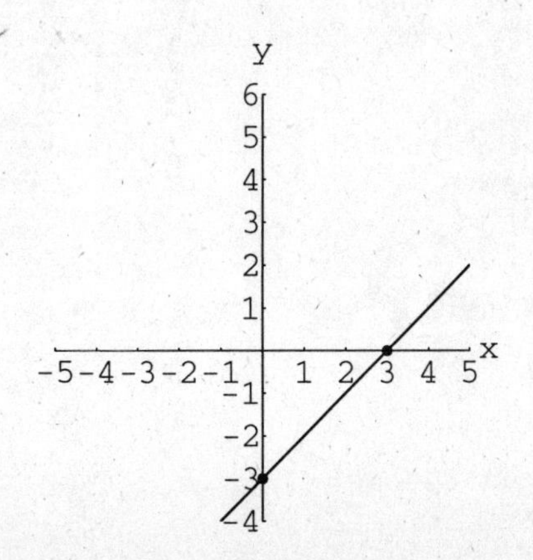

5. (a)

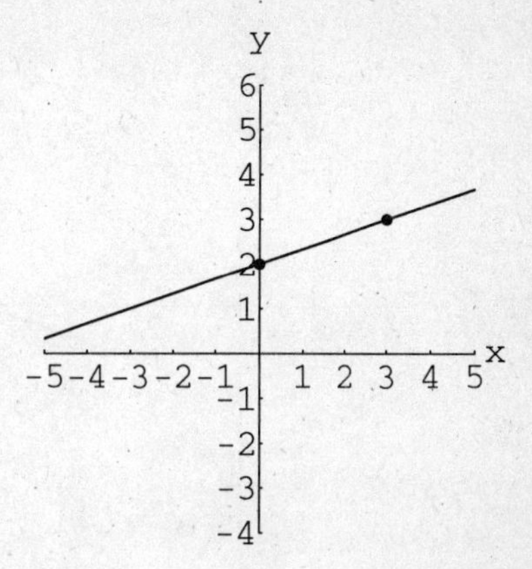

(b)

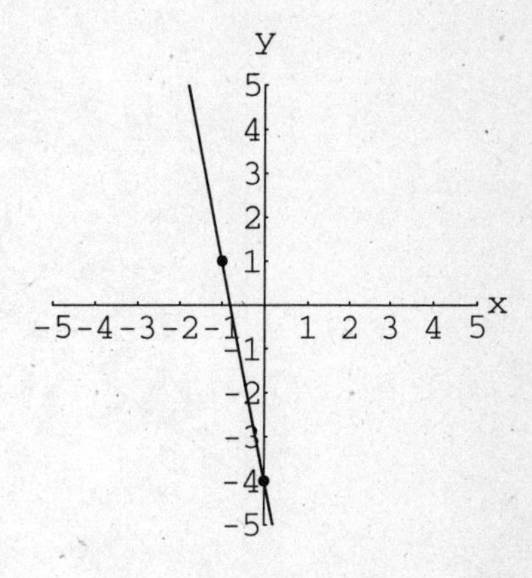

6. (a)

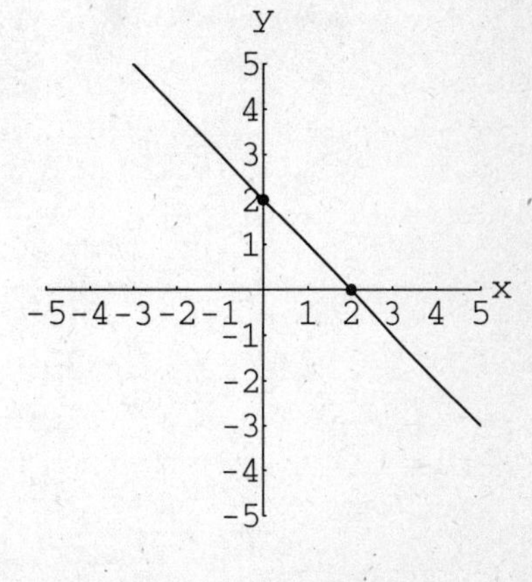

(b)

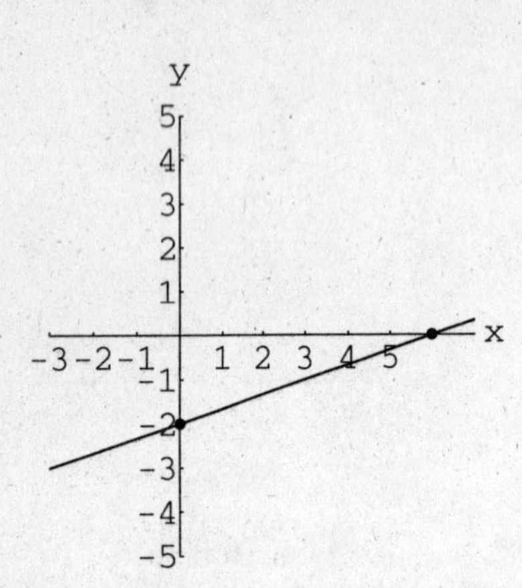

(c)

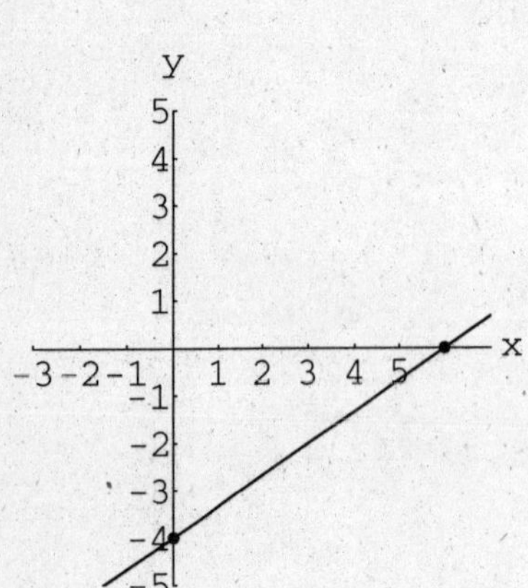

7. (a)

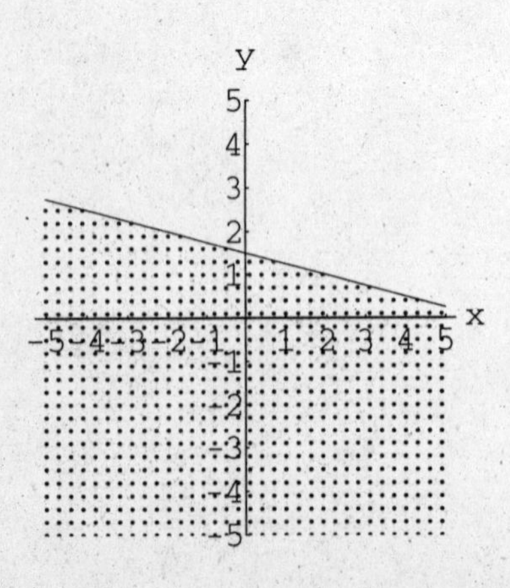

(b)

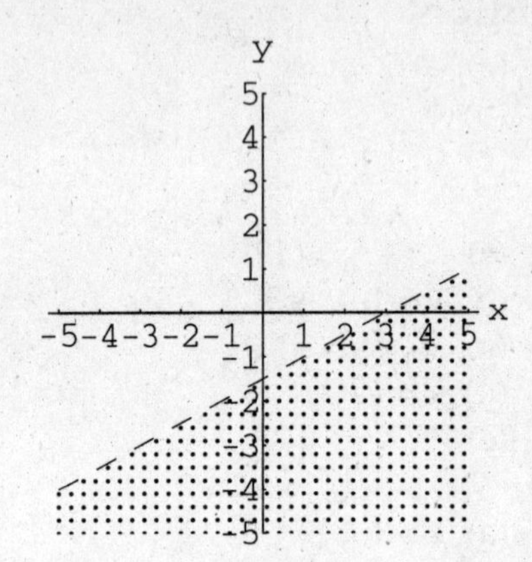

8. (a) (6, –1)
   (b) (2, 1)
   (c) (4/5, 6/5)

9. (a) (13/2, 1/2)
   (b) dependent
   (c) (2, 0)

10. (a) (7, 3, 2)
    (b) (0, 1, 2)
    (c) (–1/3, 4/3, 4)

2

Matrices

This chapter introduces matrix notation and operations. Operations such as matrix addition, subtraction, scalar multiplication, and matrix multiplication are covered. The processes involved include obtaining determinants, finding inverses, and using Cramer's Rule.

2.1 MATRIX TERMINOLOGY

> The **elements** of a matrix are the individual numbers within the matrix.

The elements are identified within a matrix by the row and column in which they are located. The rows are numbered from top to bottom, and the columns are numbered from left to right (see Figure 2.1).

$$\text{Row}\quad \begin{array}{c} \\ 1 \\ 2 \\ 3 \end{array}\overset{\text{Column}}{\begin{array}{cccc} 1 & 2 & 3 & 4 \\ \end{array}} \begin{bmatrix} 3 & 4 & -2 & 4 \\ 7 & 0 & -6 & 2 \\ 1 & 1 & 3 & -2 \end{bmatrix}$$

Figure 2.1 Matrix with Rows and Columns Identified

The dimension of a matrix is determined by the number of rows and columns it has. For example a 3×4 matix has 3 rows and 4 columns (see Figure 2.1). All of the numbers inside of a matrix are identified by first giving the row and then the column it occupies. The row and column number in the subscript are separated by a comma, but the comma is usually omitted when there are less than ten rows and less than ten columns.

Many texts use i for rows and j for columns, so that an element from matrix **A** would be identified a_{ij} or $a_{\text{row, column}}$.

In this text, every matrix operation and every element label uses the row designation first. This may help you avoid applying a column designation for a row or vice versa. A capital letter is used to identify a given matrix.

2 × 2 matrix

$$\begin{bmatrix} 2 & -4 \\ 3 & 5 \end{bmatrix}$$

1 × 1 matrix

$$\begin{bmatrix} -4 \end{bmatrix}$$

3 × 3 matrix

$$\begin{bmatrix} 2 & 1 & 0 \\ -1 & 3 & 6 \\ -2 & -2 & 4 \end{bmatrix}$$

4 × 3 matrix

$$\begin{bmatrix} 1 & -3 & 5 \\ -2 & 0 & 7 \\ 8 & 1 & -2 \\ 2 & 3 & 0 \end{bmatrix}$$

Figure 2.2 Dimensions of Matrices

EXAMPLE 2.1

For the following matrix, **A**, find the values of elements
(a) a_{12}
(b) a_{23}
(c) a_{32}

$$\mathbf{A} = \begin{bmatrix} 2 & 3 & -1 & 4 \\ 7 & 2 & -5 & 1 \\ 7 & 1 & 6 & -2 \end{bmatrix}$$

SOLUTION 2.1

(a) a_{12} is the number in the first row and second column which is 3.

(b) a_{23} is the number in the second row and third column which is –5.

(c) a_{32} is the number in the third row and second column which is 1.

2.2 OPERATIONS WITH MATRICES

Matrix Addition and Subtraction

The **addition of two matrices** is the simple addition of **corresponding elements** of the two matrices. **Subtraction** is the simple subtraction of the **corresponding elements**. Figure 2.3 shows an addition and a subtraction of matrices. Note that only matrices with the same dimensions can be added or subtracted.

$$\begin{bmatrix} 9 & -3 \\ -2 & 4 \end{bmatrix} + \begin{bmatrix} 5 & 0 \\ -2 & 5 \end{bmatrix} = \begin{bmatrix} 9+5 & -3+0 \\ -2-2 & 4+5 \end{bmatrix} = \begin{bmatrix} 14 & -3 \\ -4 & 9 \end{bmatrix}$$

$$\begin{bmatrix} 1 & 3 & 4 \end{bmatrix} - \begin{bmatrix} -2 & 1 & 5 \end{bmatrix} = \begin{bmatrix} 1-(-2) & 3-1 & 4-5 \end{bmatrix}$$
$$= \begin{bmatrix} 3 & 2 & -1 \end{bmatrix}$$

Figure 2.3 Addition and Subtraction of Matrices

Scalar Multiplication of Matrices

When we multiply a matrix by a number, we call that number a **scalar**. When a scalar is multiplied times a matrix, every element in the matrix is multiplied by the scalar number. Figure 2.4 shows a scalar multiplication.

$$4 \cdot \begin{bmatrix} 2 & -1 \\ 4 & 5 \end{bmatrix} = \begin{bmatrix} 4 \cdot 2 & 4 \cdot -1 \\ 4 \cdot 4 & 4 \cdot 5 \end{bmatrix} = \begin{bmatrix} 8 & -4 \\ 16 & 20 \end{bmatrix}$$

Figure 2.4 Scalar Multiplication

EXAMPLE 2.2

Given the following matrices:

$$\mathbf{A} = \begin{bmatrix} 5 & 0 \\ 3 & -1 \end{bmatrix} \quad \mathbf{B} = \begin{bmatrix} 1 & -3 \\ 2 & 2 \end{bmatrix} \quad \mathbf{C} = \begin{bmatrix} 4 \\ 0 \\ 5 \end{bmatrix} \quad \mathbf{D} = \begin{bmatrix} 1 \\ -2 \\ 3 \end{bmatrix}$$

Find:

(a) $2\mathbf{A}$

(b) $\mathbf{A} + \mathbf{B}$

(c) $2\mathbf{A} - 3\mathbf{B}$
(d) $\mathbf{C} - 2\mathbf{D}$
(e) $\mathbf{C} + 3\mathbf{D}$

SOLUTION 2.2

(a)

$$2\mathbf{A} = 2 \cdot \begin{bmatrix} 5 & 0 \\ 3 & -1 \end{bmatrix}$$

Write 2 times the matrix.

$$2\mathbf{A} = \begin{bmatrix} 2 \cdot 5 & 2 \cdot 0 \\ 2 \cdot 3 & 2 \cdot -1 \end{bmatrix}$$

Multiply **each** element of matrix **A** by two.

$$2\mathbf{A} = \begin{bmatrix} 10 & 0 \\ 6 & -2 \end{bmatrix}$$

The result shows that each element has been doubled.

(b)

$$\mathbf{A} = \begin{bmatrix} 5 & 0 \\ 3 & -1 \end{bmatrix} \quad \mathbf{B} = \begin{bmatrix} 1 & -3 \\ 2 & 2 \end{bmatrix}$$

Copy matrices A and B.

$$\mathbf{A} + \mathbf{B} = \begin{bmatrix} 5 & 0 \\ 3 & -1 \end{bmatrix} + \begin{bmatrix} 1 & -3 \\ 2 & 2 \end{bmatrix}$$

Write the matrices as a sum.

$$\mathbf{A} + \mathbf{B} = \begin{bmatrix} 5 + 1 & 0 - 3 \\ 3 + 2 & -1 + 2 \end{bmatrix}$$

Sum the corresponding elements.

$$\mathbf{A} + \mathbf{B} = \begin{bmatrix} 6 & -3 \\ 5 & 1 \end{bmatrix}$$

The new matrix is the final sum.

(c)

$$2\mathbf{A} = 2 \cdot \begin{bmatrix} 5 & 0 \\ 3 & -1 \end{bmatrix}$$

Copy 2 times matrix A.

$$2\mathbf{A} = \begin{bmatrix} 2 \cdot 5 & 2 \cdot 0 \\ 2 \cdot 3 & 2 \cdot -1 \end{bmatrix} = \begin{bmatrix} 10 & 0 \\ 6 & -2 \end{bmatrix}$$

Multiply each element by 2. The resulting matrix has each element doubled.

$$3\mathbf{B} = 3 \cdot \begin{bmatrix} 1 & -3 \\ 2 & 2 \end{bmatrix} = \begin{bmatrix} 3 \cdot 1 & 3 \cdot -3 \\ 3 \cdot 2 & 3 \cdot 2 \end{bmatrix}$$

Copy 3 times matrix B. Multiply each element of B by 3.

$$\mathbf{3B} = \begin{bmatrix} 3 & -9 \\ 6 & 6 \end{bmatrix}$$

The resulting matrix has each element tripled.

$$\mathbf{2A} + \mathbf{3B} = \begin{bmatrix} 10 & 0 \\ 6 & -2 \end{bmatrix} + \begin{bmatrix} 3 & -9 \\ 6 & 6 \end{bmatrix}$$

Add 2A to 3B.

$$\mathbf{2A} + \mathbf{3B} = \begin{bmatrix} 10+3 & 0-9 \\ 6+6 & -2+6 \end{bmatrix}$$

Obtain the sum and the answer.

$$= \begin{bmatrix} 13 & -9 \\ 12 & 4 \end{bmatrix}$$

(d)

$$\mathbf{2D} = 2 \cdot \begin{bmatrix} 1 \\ -2 \\ 3 \end{bmatrix} = \begin{bmatrix} 2 \cdot 1 \\ 2 \cdot -2 \\ 2 \cdot 3 \end{bmatrix} = \begin{bmatrix} 2 \\ -4 \\ 6 \end{bmatrix}$$

Write and multiply two times matrix D.

$$\mathbf{C} - \mathbf{2D} = \underset{\text{C}}{\begin{bmatrix} 4 \\ 0 \\ 5 \end{bmatrix}} - \underset{\text{2D}}{\begin{bmatrix} 2 \\ -4 \\ 6 \end{bmatrix}} = \begin{bmatrix} 4-2 \\ 0-(-4) \\ 5-6 \end{bmatrix}$$

Subtract matrix 2D from matrix C.

$$= \begin{bmatrix} 2 \\ 4 \\ -1 \end{bmatrix}$$

The solution to the problem.

(e)

$$\mathbf{3D} = 3 \cdot \begin{bmatrix} 1 \\ -2 \\ 3 \end{bmatrix} = \begin{bmatrix} 3 \cdot 1 \\ 3 \cdot -2 \\ 3 \cdot 3 \end{bmatrix} = \begin{bmatrix} 3 \\ -6 \\ 9 \end{bmatrix}$$

Write 3 times matrix D.

$$\mathbf{C} + \mathbf{3D} = \underset{\text{C}}{\begin{bmatrix} 4 \\ 0 \\ 5 \end{bmatrix}} + \underset{\text{3D}}{\begin{bmatrix} 3 \\ -6 \\ 9 \end{bmatrix}} = \begin{bmatrix} 4+3 \\ 0+-6 \\ 5+9 \end{bmatrix}$$

Add 3 times matrix D to C to find the final solution.

$$= \begin{bmatrix} 7 \\ -6 \\ 14 \end{bmatrix}$$ The solution to the problem.

Note in each of these examples that it is possible to multiply a matrix with any number or rows and columns by a scalar. Only two matrices with equal dimensions can be added or subtracted.

Matrix Multiplication

The multiplication of one matrix by another does *not* require that the matrices be the same size. The number of columns in the first matrix must equal the number of rows in the second matrix. For any two matrices **A** and **B** with rows and columns of $i \times j$ and $n \times m$ respectively, Figure 2.5 shows these requirements.

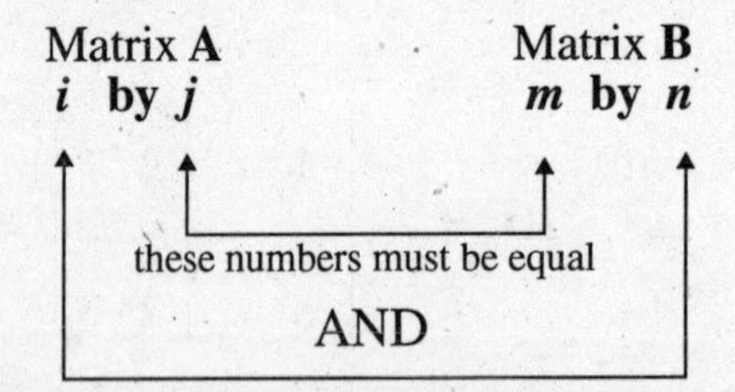

Figure 2.5 Matrix Multiplication

The product of matrix **A** (which is size ***i*** rows by ***j*** columns) and matrix **B** (which is size ***m*** rows by ***n*** columns) is a new matrix that is ***i*** rows by ***n*** columns. Therefore, a matrix multiplication of a 3 (*i* rows) by 2 (*j* columns) matrix times a 2 (*m* rows) by 4 (*n* columns) matrix forms a new matrix that is 3 (*i* rows) by 4 (*n* columns).

We are unable to multiply a 4 (*i* rows) by 2 (*j* columns) matrix times a 3 (*m* rows) by 4 (*n* columns) matrix because the number of columns in the first matrix (2) does not equal the number of rows (3) in the second matrix.

Unlike multiplication of real numbers, the product of matrices may not be the same if the matrices are multiplied in the reverse order. For example, the matrix multiplication of **AB** is not necessarily equal to **BA**.

The multiplication process can be exemplified by the two matrices **A** and **B** shown in Figure 2.6.

$$\mathbf{A} = \begin{bmatrix} 1 & 2 \\ 4 & -3 \end{bmatrix} \quad \mathbf{B} = \begin{bmatrix} -1 & 0 \\ 5 & 6 \end{bmatrix} \quad \mathbf{AB} = \begin{bmatrix} (1 \cdot -1 + 2 \cdot 5) & \\ & \end{bmatrix}$$

Figure 2.6 Step One of Example Multiplication

The first element of the top row of matrix **A** is multiplied by the first element of the left column of matrix **B**. This product is then added to the product of the second element of row 1 of matrix **A** times the bottom element of the left column of matrix **B**. (Refer to Figure 2.6.)

$$\mathbf{A} = \begin{bmatrix} 1 & 2 \\ 4 & -3 \end{bmatrix} \quad \mathbf{B} = \begin{bmatrix} -1 & 0 \\ 5 & 6 \end{bmatrix} \quad \mathbf{AB} = \begin{bmatrix} (1 \cdot -1 + 2 \cdot 5) & (1 \cdot 0 + 2 \cdot 6) \\ & \end{bmatrix}$$

Figure 2.7 Step Two of Example Multiplication

The first element of row 1 for matrix **A** is then multiplied by the top element of the right column of matrix **B**. The second element of row 1 of matrix **A** is then multiplied by the bottom element in the right-hand column of matrix **B**. (See Figure 2.7.)

$$\mathbf{A} = \begin{bmatrix} 1 & 2 \\ 4 & -3 \end{bmatrix} \quad \mathbf{B} = \begin{bmatrix} -1 & 0 \\ 5 & 6 \end{bmatrix}$$

$$\mathbf{AB} = \begin{bmatrix} (1 \cdot -1 + 2 \cdot 5) & (1 \cdot 0 + 2 \cdot 6) \\ (4 \cdot -1 + (-3) \cdot 5) & \end{bmatrix}$$

Figure 2.8 Step Three of Example Multiplication

The first element of row 2 for matrix **A** is then multiplied by the top element of the left column of matrix **B**. The second element of row 2 of matrix **A** is then multiplied by the bottom element in the left-hand column of matrix **B**. (See Figure 2.8.)

$$\mathbf{A} = \begin{bmatrix} 1 & 2 \\ 4 & -3 \end{bmatrix} \quad \mathbf{B} = \begin{bmatrix} -1 & 0 \\ 5 & 6 \end{bmatrix}$$

$$\mathbf{AB} = \begin{bmatrix} (1 \cdot -1 + 2 \cdot 5) & (1 \cdot 0 + 2 \cdot 6) \\ (4 \cdot -1 + (-3) \cdot 5) & (4 \cdot 0 + (-3) \cdot 6) \end{bmatrix}$$

$$\mathbf{A} = \begin{bmatrix} 1 & 2 \\ 4 & -3 \end{bmatrix} \quad \mathbf{B} = \begin{bmatrix} -1 & 0 \\ 5 & 6 \end{bmatrix}$$

$$\mathbf{AB} = \begin{bmatrix} (1 \cdot -1 + 2 \cdot 5) & (1 \cdot 0 + 2 \cdot 6) \\ (4 \cdot -1 + (-3) \cdot 5) & \end{bmatrix}$$

Figure 2.9 Step Four of Example Multiplication

The first element of row 2 for matrix **A** is then multiplied by the top element of the right column of matrix **B**. The second element of row 2 of matrix **A** is then multiplied by the bottom element in the right-hand column of matrix **B**. (See Figure 2.9.)

The final matrix is then simplified by adding the products from each of the previous steps (Figure 2.10.)

$$\mathbf{AB} = \begin{bmatrix} 9 & 12 \\ -19 & -18 \end{bmatrix}$$

Figure 2.10 The Product of A times B

The previous multiplication was possible because the number of rows of matrix **A** equaled the number of columns of matrix **B**.

The result was a 2×2 matrix that has the number of rows of matrix **A** and the number of columns of matrix **B**.

Note that the product of **AB** in Figure 2.10 *is not the same* as the product of **BA** in Figure 2.11.

$$\mathbf{BA} = \begin{bmatrix} -1 & -2 \\ 29 & -8 \end{bmatrix}$$

Figure 2.11 The Product of B times A

EXAMPLE 2.3

Perform the following matrix multiplications.

(a) $\begin{bmatrix} 2 \\ -1 \\ 5 \end{bmatrix} \cdot \begin{bmatrix} 0 & -3 & 2 \end{bmatrix}$

(b) $\begin{bmatrix} 1 & -3 & 2 \end{bmatrix} \cdot \begin{bmatrix} 4 \\ 7 \\ -2 \end{bmatrix}$

(c) $\begin{bmatrix} 2 \\ 3 \end{bmatrix} \cdot \begin{bmatrix} 1 & -5 \end{bmatrix}$

(d) $\begin{bmatrix} 1 & 0 \\ 3 & 2 \end{bmatrix} \cdot \begin{bmatrix} 4 & -9 & 11 \end{bmatrix}$

SOLUTION 2.3

(a) $\begin{bmatrix} \underrightarrow{2} \\ -1 \\ 5 \end{bmatrix} \cdot \begin{bmatrix} 0\downarrow & -3\downarrow & 2\downarrow \end{bmatrix} \qquad \begin{bmatrix} 2 \cdot 0 & 2 \cdot -3 & 2 \cdot 2 \\ & & \\ & & \end{bmatrix}$

The first row (one element) is multiplied by each column of the second matrix.

$\begin{bmatrix} 2 \\ \underrightarrow{-1} \\ 5 \end{bmatrix} \cdot \begin{bmatrix} 0\downarrow & -3\downarrow & 2\downarrow \end{bmatrix} \qquad \begin{bmatrix} 2 \cdot 0 & 2 \cdot -3 & 2 \cdot 2 \\ -1 \cdot 0 & -1 \cdot -3 & -1 \cdot 2 \\ & & \end{bmatrix}$

The second row (one element) is multiplied times each column of the second matrix.

$\begin{bmatrix} 2 \\ -1 \\ \underrightarrow{5} \end{bmatrix} \cdot \begin{bmatrix} 0\downarrow & -3\downarrow & 2\downarrow \end{bmatrix} \qquad \begin{bmatrix} 2 \cdot 0 & 2 \cdot -3 & 2 \cdot 2 \\ -1 \cdot 0 & -1 \cdot -3 & -1 \cdot 2 \\ 5 \cdot 0 & 5 \cdot -3 & 5 \cdot 2 \end{bmatrix}$

The third row (one element) is multiplied by each column.

$$\begin{bmatrix} 0 & -6 & 4 \\ 0 & 3 & -2 \\ 0 & -15 & 10 \end{bmatrix}$$

The result is a 3×3 matrix.

(b) $\begin{bmatrix} 1 & -3 & 2 \end{bmatrix} \cdot \begin{bmatrix} 4 \\ 7 \\ -2 \end{bmatrix} = \begin{bmatrix} 1 \cdot 4 + (-3) \cdot 7 + 2 \cdot (-2) \end{bmatrix}$

The only row of the first matrix will be multiplied by the only column of the second matrix. The result is a 1×1 matrix.

Since $4 + (-21) + (-4) = -17 + (-4) = -21$, the resulting matrix is $\begin{bmatrix} -21 \end{bmatrix}$.

(c) $\begin{bmatrix} 2 \\ 3 \end{bmatrix} \cdot \begin{bmatrix} 1 & -5 \end{bmatrix} = \begin{bmatrix} 2 \cdot 1 & 2 \cdot -5 \\ & \end{bmatrix}$

The first element in the top row is multiplied by the two columns of the second matrix.

$$\begin{bmatrix} 2 \\ 3 \end{bmatrix} \cdot \begin{bmatrix} 1 & -5 \end{bmatrix} = \begin{bmatrix} 2 \cdot 1 & 2 \cdot -5 \\ 3 \cdot 1 & 3 \cdot -5 \end{bmatrix}$$

The element of the second row is taken times the columns of the second matrix. The result is a 2×2 matrix.

$$= \begin{bmatrix} 2 & -10 \\ 3 & -15 \end{bmatrix}$$

(d) these are not equal

2×2 $\qquad$ 1×3

$$\begin{bmatrix} 1 & 0 \\ 3 & 2 \end{bmatrix} \qquad \begin{bmatrix} 4 & -9 & 11 \end{bmatrix}$$

These matrices can't be multiplied since the number of columns in the first matrix does not equal the number of rows in the second matrix.

2.3 FINDING AND USING DETERMINANTS OF A SQUARE MATRIX

The solution to a system of equations can be found through a method that determines a single number representative of all the elements in a particular matrix. This single numerical value is called the **determinant**.

The determinant of matrix **A** is indicated by the abbreviation det **A**, or by writing the elements of the matrix within vertical bars instead of brackets.

Finding the Determinant of a 2 x 2 (Order 2) Matrix

A 2 × 2 matrix is a square matrix containing four elements. We can find the determinant of matrix **A** by computing $ad - bc$. That is, if

$\mathbf{A} = \begin{bmatrix} a & b \\ c & d \end{bmatrix}$, then det $\mathbf{A} = ad - bc$. For example, if $\mathbf{A} = \begin{bmatrix} 2 & -1 \\ 3 & 4 \end{bmatrix}$, then

$$\det \mathbf{A} = \begin{vmatrix} 2 & -1 \\ 3 & 4 \end{vmatrix} = ad - bc = (2)(4) - (-1)(3) = 8 - (-3) = 8 + 3 = 11.$$

Note that we use straight lines to imply that a determinant is needed as opposed to the brackets used to enclose a matrix.

This formula for finding a determinant can be visualized as multiplication along a diagonal. Given a 2 × 2 matrix, $\mathbf{B} = \begin{bmatrix} 2 & -3 \\ 0 & 5 \end{bmatrix}$, we multiply *down* a diagonal from the upper left to lower right.

$$\det \mathbf{B} = \begin{vmatrix} 2 & -3 \\ 0 & 5 \end{vmatrix}$$

(2)(5) = 10

This product is called the bottom product. We also multiply up the diagonal from lower left to upper right.

$$\det \mathbf{B} = \begin{vmatrix} 2 & -3 \\ 0 & 5 \end{vmatrix}$$

(0)(–3) = 0
Top

(2)(5) = 10
Bottom

The second product is the *top* product. The subtraction of the *top* from the *bottom* gives the determinant.

10	−	0	=	10	=	det **B**
bottom	−	top				

Finding the Determinant of a 3 x 3 (Order 3) Matrix

Larger square matrices, such as a 3×3 matrix, require a somewhat different method for finding a determinant. For a given matrix, **A**, the formula for finding the determinant is given below.

$$\text{If } A = \begin{bmatrix} a_{11} & a_{12} & a_{13} \\ a_{21} & a_{22} & a_{23} \\ a_{31} & a_{32} & a_{33} \end{bmatrix}, \text{ then } \det \mathbf{A} = a_{11}(a_{22}a_{33} - a_{32}a_{23}) - a_{21}(a_{12}a_{33} - a_{32}a_{13}) + a_{31}(a_{12}a_{23} - a_{22}a_{13})$$

This formula appears to be very complicated, but it can be simplified by using *minors* and *cofactors*.

> A **minor** of an element in a 3×3 matrix is the 2×2 matrix obtained by eliminating the row and column in which the element resides.

For example, the minor of a_{11} is $\begin{bmatrix} a_{22} & a_{23} \\ a_{32} & a_{33} \end{bmatrix}$ which is determined by eliminating the row and column in which a_{11} resides.

$$\begin{bmatrix} a_{11} & a_{12} & a_{13} \\ a_{21} & a_{22} & a_{23} \\ a_{31} & a_{32} & a_{33} \end{bmatrix}$$

minor

> The *cofactor* of an element, a_{ij}, is $(-1)^{i+j}$ times the determinant of the minor of a_{ij}.

In our matrix, **A**, the *cofactor* of a_{11} is $(-1)^{1+1}\begin{vmatrix} a_{22} & a_{23} \\ a_{32} & a_{33} \end{vmatrix}$ or

$+1(a_{22}\,a_{33} + a_{32}\,a_{23})$ and the cofactor for a_{21} is $(-1)^{2+1}\begin{vmatrix} a_{22} & a_{23} \\ a_{32} & a_{33} \end{vmatrix}$

or $-1(a_{12}\,a_{33} - a_{32}\,a_{13})$.

The Diagonal Expansion Formula for a 3 x 3 Matrix

A more mechanical technique for finding the determinant of a 3 × 3 matrix is the diagonal expansion formula. We expand a 3 × 3 matrix into a 3 × 5 matrix by adding copies of columns 1 and 2 to the right side of the 3 × 3 matrix. Then we multiply down (or up) diagonals to obtain an upper and lower product-sum.

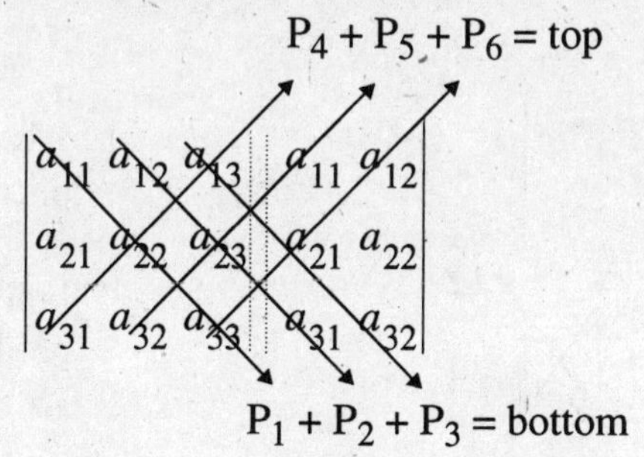

det **A** = bottom – top

EXAMPLE 2.4

Find the determinant of the following matrices.

(a) $\mathbf{A} = \begin{bmatrix} 3 & 5 \\ 7 & -1 \end{bmatrix}$

(b) $\mathbf{B} = \begin{bmatrix} 4 & 2 \\ 6 & 3 \end{bmatrix}$

(c) $\mathbf{C} = \begin{bmatrix} 4 & 1 & 0 \\ 3 & -1 & 2 \\ 0 & 5 & 6 \end{bmatrix}$

(d) $\mathbf{D} = \begin{bmatrix} 3 & -2 & 1 \\ 0 & -1 & 6 \\ 0 & 4 & -3 \end{bmatrix}$

SOLUTION 2.4

(a) $\det \mathbf{A} = \begin{vmatrix} 3 & 5 \\ 7 & -1 \end{vmatrix}$

We will use the formula: $ad - bc$. Let $3 = a$, $5 = b$, $7 = c$, $-1 = d$.

$$\det \mathbf{A} = ad - bc$$
$$= (3)(-1) - (7)(5)$$
$$= -3 - 35 = -38$$

Use the formula to find the determinant.

(b) $\det \mathbf{B} = \begin{vmatrix} 4 & 2 \\ 6 & 3 \end{vmatrix}$

We will use diagonal multiplication.

$$\det \mathbf{B} = \begin{vmatrix} 4 & 2 \\ 6 & 3 \end{vmatrix} \quad \begin{matrix} \nearrow 12 \\ \searrow 12 \end{matrix}$$

Find the two diagonal products for top and bottom.

$$\det \mathbf{B} = \text{bottom} - \text{top}$$
$$= 12 - 12 = 0$$

Subtract top from bottom to find the determinant.

(c) $\mathbf{C} = \begin{bmatrix} 4 & 1 & 0 \\ 3 & -1 & 2 \\ 0 & 5 & 6 \end{bmatrix}$

We will use the cofactor method to find the determinant.

$$\mathbf{C} = \begin{bmatrix} \boxed{4} & 1 & 0 \\ \boxed{3} & -1 & 2 \\ \boxed{0} & 5 & 6 \end{bmatrix}$$

We can choose any row or column as the primary elements. We choose the first column.

$$\det \mathbf{C} = 4\,(-1)^{1+1} \begin{vmatrix} -1 & 2 \\ 5 & 6 \end{vmatrix}$$

Write the elements of the first column with each of their cofactors. Add all the products.

$$+\,3(-1)^{2+1}\begin{vmatrix}1 & 0\\ 5 & 6\end{vmatrix}$$

$$+\,0(-1)^{3+1}\begin{vmatrix}1 & 0\\ -1 & 2\end{vmatrix}$$

$\det \mathbf{C} = 4(1)(-6-10) + (3)(-1)(6-0) + (0)(1)(2+0)$ — Find the diagonal product of each 2×2 matrix.

$= 4(-16) - 3(6) + 0$

$\det \mathbf{C} = -64 - 18 = -82$ — Complete the multiplication and addition.

(d) $\mathbf{D} = \begin{bmatrix}3 & -2 & 1\\ 0 & -1 & 6\\ 0 & 4 & -3\end{bmatrix}$ — We will use the diagonal expansion formula.

$$\begin{vmatrix}3 & -2 & 1 & 3 & -2\\ 0 & -1 & 6 & 0 & -1\\ 0 & 4 & -3 & 0 & 4\end{vmatrix}$$

Append the first two columns to the right of the matrix.

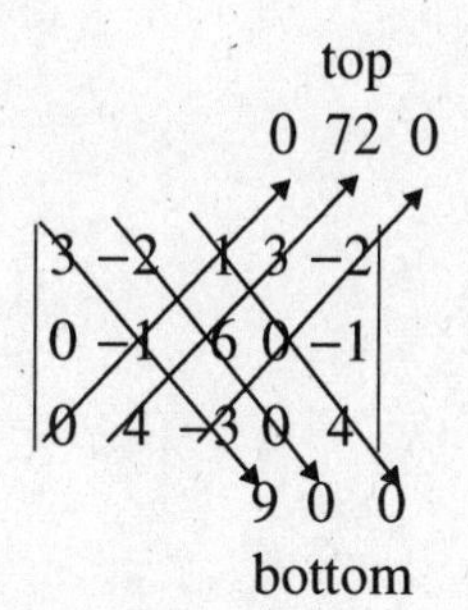

Find the diagonal products.

$\det \mathbf{D} = (9 + 0 + 0) - (0 + 72 + 0)$

$= (9 - 72) = -63$

Add the bottom products and the top products. Subtract the top sum from the bottom sum.

2.4 USING CRAMER'S RULE

Cramer's Rule is a technique for solving systems of equations using determinants. Before the required determinants are found, we must place some of the coefficients of the equations in a form called an **augmented matrix**. An augmented matrix includes some of the coefficients for the variables, as well as the constants to which the variable terms are equal. A determinant can not be found for any matrix other than a **square matrix**. A **square matrix** has the same number of rows as columns. We convert the augmented matrix into a square matrix by deleting the appropriate column. This conversion process, and the solution to a system of equations, follows.

Finding the Solution to an Augmented Matrix

$$\begin{aligned} x+y+z&=6 \\ 2x-y+3z&=9 \\ x+y-z&=0 \end{aligned}$$

A system of three equations can be converted to matrix form. When the coefficients and constants are written in matrix form, it is called **the augmented matrix form**. We do not use the entire augmented matrix to form a determinant.

$$\left|\begin{array}{ccc:c} 1 & 1 & 1 & 6 \\ 2 & -1 & 3 & 9 \\ 1 & 1 & -1 & 0 \end{array}\right|$$

The matrix must be a "square" matrix to form a determinant. A square matrix has the same numbers of rows and columns. This matrix is 3×4, so we remove the last column to find the determinant.

$$\det \mathbf{D} = \begin{vmatrix} 1 & 1 & 1 \\ 2 & -1 & 3 \\ 1 & 1 & -1 \end{vmatrix}$$

$$= 6 - 0 = \textcircled{6}$$

The determinant of matrix **D** is found. It will be used as a **denominator** for three fractions that will give the solutions to x, y, and z.

$$\det \boldsymbol{x} = \begin{vmatrix} 6 & 1 & 1 \\ 9 & -1 & 3 \\ 0 & 1 & -1 \end{vmatrix}$$

$$= 15 - 9 = \textcircled{6}$$

To find the **numerator** for the $\boldsymbol{x}$-value, we replace the first column with the fourth column. We then find the determinant.

$$\det \boldsymbol{y} = \begin{vmatrix} 1 & 6 & 1 \\ 2 & 9 & 3 \\ 1 & 0 & -1 \end{vmatrix}$$

$$= 9 - (-3) = \textcircled{12}$$

To find the **numerator** for the $\boldsymbol{y}$-value, we replace the second column with the fourth column. We then find the determinant.

$$\det z = \begin{vmatrix} 1 & 1 & 6 \\ 2 & -1 & 9 \\ 1 & 1 & 0 \end{vmatrix}$$

$$= 21 - 3 = \textcircled{18}$$

To find the **numerator** for the z-value, we replace the third column with the fourth column. We then find the determinant.

The value for $x = \dfrac{\text{det.}\,\boldsymbol{x}}{\text{det. D}} = \dfrac{6}{6} = 1.$

The value for $y = \dfrac{\text{det.}\,\boldsymbol{y}}{\text{det. D}} = \dfrac{12}{6} = 2.$

The value for $z = \dfrac{\text{det.}\,z}{\text{det. D}} = \dfrac{18}{6} = 3.$

The solution for this system is (1, 2, 3). As you can see, the use of determinants is a rather lengthy process. While this technique is easier to demonstrate in smaller systems, it is usually reserved for use on computers with much larger systems of equations.

EXAMPLE 2.5

Solve the following systems of equations through the use of matrices and determinants.

(a) $2x - y = 5$
$3x + y = 10$

(b)
$$x - y - z = 4$$
$$2x - 5y - z = 6$$
$$x + y + 2z = 9$$

SOLUTION 2.5

(a)
$$2x - y = 5$$
$$3x + y = 10$$

Copy the system of equations.

$$\left|\begin{array}{cc:c} 2 & -1 & 5 \\ 3 & 1 & 10 \end{array}\right|$$

Form the augmented matrix.

$$\det D = \begin{vmatrix} 2 & -1 \\ 3 & 1 \end{vmatrix}$$
$$= 2 - -3 = \textcircled{5}$$

The determinant for the **denominator** is 5.

$$\det X = \begin{vmatrix} 5 & -1 \\ 10 & 1 \end{vmatrix}$$
$$= 5 - (-10) = \textcircled{15}$$

The augmented column replaces the ***x*** -coefficients. The determinant that is the **numerator** for the ***x***-value is 15.

$$\det Y = \begin{vmatrix} 2 & 5 \\ 3 & 10 \end{vmatrix}$$
$$= 20 - 15 = \textcircled{5}$$

The augmented column replaces the ***y***-coefficients. The determinant that is the **numerator** for the ***y***-value is 5.

The ***x***-value is 15/5 = **3**. The ***y***-value is 5/5 = **1**. The solution set is **(3, 1)** for this system of equations.

(b)
$$x - y - z = 4$$
$$2x - 5y - z = 6$$
$$x + y + 2z = 9$$

This system of three equations can be converted to matrix form. When the coefficients and constants are written in matrix form, it is called **an augmented matrix form**.

$$\begin{vmatrix} 1 & -1 & -1 & 4 \\ 2 & -5 & -1 & 6 \\ 1 & 1 & 2 & 9 \end{vmatrix}$$

You can only use square submatrices of the augumented matrix. This matrix is 3 × 4, so we remove the last column to find the determinant.

$$\det = \mathbf{D} = \begin{vmatrix} 1 & -1 & -1 \\ 2 & -5 & -1 \\ 1 & 1 & 2 \end{vmatrix}$$

$$= -11 - 0 = -11$$

The determinant of matrix **D** is found. It will be used as a **denominator** for the three fractions that will give the answers to *x*, *y*, and *z*.

$$\det x = \begin{vmatrix} 4 & 1 & -1 \\ 6 & -5 & 1 \\ 9 & 1 & 2 \end{vmatrix}$$

$$= -37 - 29 = -66$$

To find the **numerator** for the ***x*** value, we replace the *x*-coefficient column with the fourth column in the augmented matrix. We then find the determinant.

$$\det y = \begin{vmatrix} 1 & 4 & -1 \\ 2 & 6 & 1 \\ 1 & 9 & 2 \end{vmatrix}$$

$$= 10 - 1 = -11$$

To find the **numerator** for the ***y*** value, we replace the *y*-coefficient column with the fourth column in the augumented matrix. We then find the determinant.

$$\det z = \begin{vmatrix} 1 & 1 & 4 \\ 2 & -5 & 6 \\ 1 & 1 & 9 \end{vmatrix}$$

$$= -43 + 32 = -11$$

To find the **numerator** for the ***z*** value, we replace the *z*-coefficient column with the fourth column in the augumented matrix. We then find the determinant.

$$\text{The value for } x = \frac{\det. x}{\det. \text{D}} = \frac{-66}{-11} = 6.$$

$$\text{The value for } y = \frac{\det. y}{\det. \text{D}} = \frac{-11}{-11} = 1.$$

The value for $z = \dfrac{\text{det. } z}{\text{det. D}} = \dfrac{-11}{-11} = 1$.

The solution for this system is (**6, 1, 1**). Check the answer on your own.

2.5 FINDING INVERSES

Matrix operations do not have a strict equivalent to division. Therefore, we recall that the multiplicative inverse is an alternative definition of division in real number arithmetic and is shown in the following expression.

$$\frac{1}{a} \cdot a = 1$$

or

$$a^{-1} \cdot a = 1$$

For a square matrix, A, the inverse is designated as A^{-1}. Therefore, if we cannot truly "divide" by A, we can perform an equivalent operation by multiplying by A^{-1}.

We can, then, simulate the division of some matrices: $B/A = C$ by using the inverse A^{-1} in $B \cdot A^{-1} = C$ (where B, A^{-1}, and C are square matrices).

We will use a given matrix A, and the identity matrix, I, to find the inverse of matrix A, A^{-1}, when it exists.

The **identity** matrix, I, is a matrix that has 1's along its main diagonal and 0's elsewhere. The determinant of an identity matrix is 1.

The identity matrix is a square matrix unique to each dimension.

$$I_2 = \begin{bmatrix} 1 & 0 \\ 0 & 1 \end{bmatrix} \qquad I_3 = \begin{bmatrix} 1 & 0 & 0 \\ 0 & 1 & 0 \\ 0 & 0 & 1 \end{bmatrix} \qquad \text{and} \qquad I_4 = \begin{bmatrix} 1 & 0 & 0 & 0 \\ 0 & 1 & 0 & 0 \\ 0 & 0 & 1 & 0 \\ 0 & 0 & 0 & 1 \end{bmatrix}$$

We will keep in mind that $A \cdot A^{-1} = I$ as we search for the inverse of matrix A if it exists.

Finding the Inverse of a 2 x 2 Matrix

There are two methods that are typically used to find the inverse of a 2 × 2 matrix. One method uses a formula, while the other uses row addition and multiplication of matrices.

Use of a Formula

For a 2 × 2 matrix labelled $\begin{bmatrix} a & b \\ c & d \end{bmatrix}$, the inverse can be found by using the formula $\left(\frac{1}{ad-bc}\right)\begin{bmatrix} d & -b \\ -c & a \end{bmatrix}$.

Note that $ad - bc$ is the determinant of a 2 × 2 matrix (see Section 2.4). If the determinant of a matrix is zero (0), then the inverse of that matrix does not exist. (Division by 0 in $1/(ad-bc)$ is not defined).

Given the matrix $\begin{bmatrix} 2 & 1 \\ 0 & 1 \end{bmatrix}$, then, the use of the formula gives us

$$\frac{1}{2(1)-0(1)}\begin{bmatrix} 1 & -1 \\ -0 & 2 \end{bmatrix} = \frac{1}{2}\begin{bmatrix} 1 & -1 \\ 0 & 2 \end{bmatrix} = \begin{bmatrix} \frac{1}{2} & -\frac{1}{2} \\ 0 & 1 \end{bmatrix} \text{ as the inverse.}$$

The matrix multiplication

$$\begin{bmatrix} 2 & 1 \\ 0 & 1 \end{bmatrix} \cdot \begin{bmatrix} \frac{1}{2} & -\frac{1}{2} \\ 0 & 1 \end{bmatrix} = \begin{bmatrix} 2\left(\frac{1}{2}\right)+0(0) & 2\left(-\frac{1}{2}\right)+(1)(1) \\ (0)\left(\frac{1}{2}\right)+(1)(0) & 0\left(-\frac{1}{2}\right)+(1)(1) \end{bmatrix}$$

$= \begin{bmatrix} 1 & 0 \\ 0 & 1 \end{bmatrix}$ the I_2 matrix.

This confirms we have obtained the inverse matrix.

Using Row Operations

We can find the inverse of a square matrix by augmenting the original matrix with the identity matrix of the same dimensions. The original matrix is then converted to the identity form by row operations. In the process, the identity matrix is converted to the inverse of the original matrix (if the inverse exists).

Symbolically, $[A|I]$ can be converted to $[I|A^{-1}]$. The process will be demonstrated through example.

Our task is to find the inverse of $\begin{bmatrix} 3 & 1 \\ 4 & 1 \end{bmatrix}$. The operation begins whe.. we augment the matrix with the I_2 matrix. Our objective is to manipulate rows and columns to convert the original matrix on the left side into I_2.

$$\begin{matrix} R_1 \rightarrow \\ R_2 \rightarrow \end{matrix} \left[\begin{array}{cc|cc} 3 & 1 & 1 & 0 \\ 4 & 1 & 0 & 1 \end{array}\right]$$

The top row is labeled R_1, and the bottom row R_2.

$$\left[\begin{array}{cc|cc} -1 & 0 & 1 & -1 \\ 4 & 1 & 0 & 1 \end{array}\right]$$

Subtract row 2 from row 1. $(R_1 - R_2)$. Place result in row 1.

$$\left[\begin{array}{cc|cc} -1 & 0 & 1 & -1 \\ 0 & 1 & 4 & -3 \end{array}\right]$$

Multiply row 1 times four and add row 2. Place the new sum in row 2.

$$\left[\begin{array}{cc|cc} 1 & 0 & -1 & 1 \\ 0 & 1 & 4 & -3 \end{array}\right]$$

Negate row 1.

We now have the augmented matrix in the form of $[I|A^{-1}]$. There fore, $A^{-1} = \begin{bmatrix} -1 & 1 \\ 4 & -3 \end{bmatrix}$ and its product with A will result in I_2.

Finding the Inverse of a 3 x 3 Matrix

The process by which row operations are used to find the inverse for a 2×2 matrix can be extended to a higher order matrix. The 3×3 matrix can be inverted with the use of the I_3 matrix.

We can use this method to find the inverse of $A = \begin{bmatrix} -1 & 0 & 2 \\ 3 & 1 & 0 \\ 1 & 1 & 1 \end{bmatrix}$ (if it exists). We begin by augmenting $[A|I_3]$ and solving for $\left[I_3|A^{-1}\right]$.

$$\begin{matrix} R_1 \rightarrow \\ R_2 \rightarrow \\ R_3 \rightarrow \end{matrix} \left[\begin{array}{ccc|ccc} -1 & 0 & 2 & 1 & 0 & 0 \\ 3 & 1 & 0 & 0 & 1 & 0 \\ 1 & 1 & 1 & 0 & 0 & 1 \end{array}\right]$$

We can label each row. R_1 is the top row, R_2 is the middle row, and R_3 is the bottom row.

$$\left[\begin{array}{ccc:ccc} 1 & 0 & -2 & -1 & 0 & 0 \\ 3 & 1 & 0 & 0 & 1 & 0 \\ 1 & 1 & 1 & 0 & 0 & 1 \end{array}\right]$$

The first row is negated.

$$\left[\begin{array}{ccc:ccc} 3 & 2 & 0 & -1 & 0 & 2 \\ 3 & 1 & 0 & 0 & 1 & 0 \\ 1 & 1 & 1 & 0 & 0 & 1 \end{array}\right] \begin{array}{l} \leftarrow 2R_3 + R_1 \\ \\ \\ \end{array}$$

$$\left[\begin{array}{ccc:ccc} 3 & 2 & 0 & -1 & 0 & 2 \\ 0 & 1 & 0 & -1 & -1 & 2 \\ 1 & 1 & 1 & 0 & 0 & 1 \end{array}\right] \begin{array}{l} \\ \leftarrow R_1 - R_2 \\ \\ \end{array}$$

$$\left[\begin{array}{ccc:ccc} 3 & 2 & 0 & -1 & 0 & 2 \\ 0 & 1 & 0 & -1 & -1 & 2 \\ 0 & -1 & -3 & -1 & 0 & -1 \end{array}\right] \begin{array}{l} \\ \\ \leftarrow R_1 - 3R_3 \end{array}$$

$$\left[\begin{array}{ccc:ccc} 3 & 2 & 0 & -1 & 0 & 2 \\ 0 & 1 & 0 & -1 & -1 & 2 \\ 0 & 0 & -3 & -2 & -1 & -1 \end{array}\right] \begin{array}{l} \\ \\ \leftarrow R_2 + R_3 \end{array}$$

$$\left[\begin{array}{ccc:ccc} 3 & 0 & 0 & 1 & 2 & -2 \\ 0 & 1 & 0 & -1 & -1 & 2 \\ 0 & 0 & -3 & -2 & -1 & 1 \end{array}\right] \begin{array}{l} \leftarrow R_1 - 2R_2 \\ \\ \\ \end{array}$$

$$\left[\begin{array}{ccc:ccc} 1 & 0 & 0 & \frac{1}{3} & \frac{2}{3} & -\frac{2}{3} \\ 0 & 1 & 0 & -1 & -1 & 2 \\ 0 & 0 & 1 & \frac{2}{3} & \frac{1}{3} & \frac{1}{3} \end{array}\right] \begin{array}{l} \frac{1}{3}R_1 \\ \\ -\frac{1}{3}R_3 \end{array}$$

$$\begin{bmatrix} \frac{1}{3} & \frac{2}{3} & -\frac{2}{3} \\ -1 & -1 & 2 \\ \frac{2}{3} & \frac{1}{3} & \frac{1}{3} \end{bmatrix}$$ is A^{-1}, the inverse of A.

During the process of finding the inverse, should any row left of the vertical line become all zeroes, then the matrix *does not have an inverse*. It is not invertible. The inverse of the matrix is unique, in that it is the only inverse of the matrix.

EXAMPLE 2.6

Find the inverse of the following matrices, if possible.

(a) $A = \begin{bmatrix} 1 & 2 \\ 0 & 4 \end{bmatrix}$

(b) $B = \begin{bmatrix} -1 & 3 \\ 2 & 1 \end{bmatrix}$

(c) $C = \begin{bmatrix} 2 & 0 & 1 \\ 1 & -1 & 2 \\ 4 & 2 & -2 \end{bmatrix}$

SOLUTION 2.6

(a) $A = \begin{bmatrix} 1 & 2 \\ 0 & 4 \end{bmatrix}$

In this example, we will use the formula

$$\frac{1}{ad - bc}\begin{bmatrix} d & -b \\ -c & a \end{bmatrix}$$

$$A^{-1} = \frac{1}{(1)\,(4) - (0)\,(2)}\begin{bmatrix} 4 & -2 \\ 0 & 1 \end{bmatrix}$$

Let $a = 1$, $b = 2$, $c = 0$, and $d = 4$.

$$A^{-1} = \frac{1}{4}\begin{bmatrix} 4 & -2 \\ 0 & 1 \end{bmatrix}$$

Complete the solution process by multiplying by the scalar, $1/4$.

$$A^{-1} = \begin{bmatrix} \frac{4}{4} & -\frac{2}{4} \\ \frac{0}{4} & \frac{1}{4} \end{bmatrix}$$

$$A^{-1} = \begin{bmatrix} 1 & -\frac{1}{2} \\ 0 & \frac{1}{4} \end{bmatrix}$$

Simplify the fractions to complete the inverse.

b) $B = \begin{bmatrix} -1 & 3 \\ 2 & 1 \end{bmatrix}$

We will use row operations to solve for this inverse.

$$\begin{array}{l} R_1 \text{—>} \\ R_2 \text{—>} \end{array} \left[\begin{array}{cc:cc} -1 & 3 & 1 & 0 \\ 2 & 1 & 0 & 1 \end{array}\right]$$

Augment the matrix, B, by appending the identity matrix, I_2.

$$\begin{array}{lrrrr} R_1 & -1 & 3 & 1 & 0 \\ -R_3 & \underline{-6} & \underline{-3} & \underline{0} & \underline{-3} \\ & -7 & 0 & 1 & -3 \end{array}$$

Replace the top row with the subraction of three times **row two** from **row one.**

$$\left[\begin{array}{cc:cc} -7 & 0 & 1 & -3 \\ 2 & 1 & 0 & 1 \end{array}\right] \text{<—} R_1 - 3R_3$$

The result replaces row one.

$$\left[\begin{array}{cc:cc} -1 & 0 & \frac{1}{7} & -\frac{3}{7} \\ 2 & 1 & 0 & 1 \end{array}\right]$$

The top row is multiplied by 1/7 to simplify row operations.

$$\begin{array}{rrrr} -2 & 0 & \frac{2}{7} & -\frac{6}{7} \\ \underline{2} & \underline{1} & \underline{0} & \underline{1} \text{ <—} R_2 \\ 0 & 1 & \frac{2}{7} & \frac{1}{7} \end{array}$$

Add two times **row one** to **row two.**

$$\left[\begin{array}{cc:cc} -1 & 0 & \frac{1}{7} & -\frac{3}{7} \\ 0 & 1 & \frac{2}{7} & \frac{1}{7} \end{array}\right]$$

The result replaces row 2.

$$\left[\begin{array}{cc|cc} 1 & 0 & -\frac{1}{7} & \frac{3}{7} \\ 0 & 1 & \frac{2}{7} & \frac{1}{7} \end{array}\right]$$

Negate row one to find the inverse, B^{-1}.

$$B^{-1} = \begin{bmatrix} -\frac{1}{7} & \frac{3}{7} \\ \frac{2}{7} & \frac{1}{7} \end{bmatrix}$$

(c) $C = \begin{bmatrix} 2 & 0 & 1 \\ 1 & -1 & 2 \\ 4 & 2 & -2 \end{bmatrix}$

We will use row operations to invert this matrix.

$$\begin{array}{l} R_1 \longrightarrow \\ R_2 \longrightarrow \\ R_3 \longrightarrow \end{array} \left[\begin{array}{ccc|ccc} 2 & 0 & 1 & 1 & 0 & 0 \\ 1 & -1 & 2 & 0 & 1 & 0 \\ 4 & 2 & -2 & 0 & 0 & 1 \end{array}\right]$$

Augment the matrix by combining with the identity matrix, I_3.

$$\begin{array}{rrrrrrr} -2R_1 \longrightarrow & -4 & 0 & -2 & -2 & 0 & 0 \\ R_2 \longrightarrow & 1 & -1 & 2 & 0 & 1 & 0 \\ \hline & -3 & -1 & 0 & -2 & 1 & 0 \end{array}$$

Add –2 times row one to row two.

$$\left[\begin{array}{ccc|ccc} -3 & -1 & 0 & -2 & 1 & 0 \\ 1 & -1 & 2 & 0 & 1 & 0 \\ 4 & 2 & -2 & 0 & 0 & 1 \end{array}\right]$$

Replace the first row with the new sum.

$$\begin{array}{rrrrrrr} R_2 \longrightarrow & 1 & -1 & 2 & 0 & 1 & 0 \\ R_3 \longrightarrow & 4 & 2 & -2 & 0 & 1 & 0 \\ \hline & 5 & 1 & 0 & 0 & 2 & 0 \end{array}$$

Add row two with row three.

$$\left[\begin{array}{ccc|ccc} -3 & -1 & 0 & -2 & 1 & 0 \\ 5 & 1 & 0 & 0 & 2 & 0 \\ 4 & 2 & -2 & 0 & 0 & 1 \end{array}\right]$$

Replace row two with the result.

$$\begin{array}{rrrrrrr} R_1 \longrightarrow & -3 & -1 & 0 & -2 & 1 & 0 \\ R_2 \longrightarrow & 5 & 1 & 0 & 0 & 2 & 0 \\ \hline & 2 & 0 & 0 & -2 & 3 & 0 \end{array}$$

Add row one to row two.

$$R_1 + R_2 \longrightarrow \left[\begin{array}{rrr:rrr} 2 & 0 & 0 & -2 & 3 & 0 \\ 5 & 1 & 0 & 0 & 2 & 0 \\ 4 & 2 & -2 & 0 & 0 & 1 \end{array}\right]$$

Replace row one with the sum.

$$\begin{array}{rrrrrrr} -2R_1 \longrightarrow & -4 & 0 & 0 & 4 & -6 & 0 \\ R_3 \longrightarrow & 4 & 2 & -2 & 0 & 0 & 1 \\ \hline & 0 & 2 & -2 & 4 & -6 & 1 \end{array}$$

Add –2 times row one to row three.

$$\begin{array}{r} \\ \\ R_3 - 2R_1 \longrightarrow \end{array} \left[\begin{array}{rrr:rrr} 2 & 0 & 0 & -2 & 3 & 0 \\ 5 & 1 & 0 & 0 & 2 & 0 \\ 0 & 2 & -2 & 4 & -6 & 1 \end{array}\right]$$

Replace row three with the results.

$$\begin{array}{rrrrrrr} -5R_1 \longrightarrow & -10 & 0 & 0 & 10 & -15 & 0 \\ 2R_2 \longrightarrow & 10 & 2 & 0 & 0 & 4 & 0 \\ \hline & 0 & 2 & 0 & 10 & -11 & 0 \end{array}$$

Add –5 times row one to 2 times row two.

$$\left[\begin{array}{rrr:rrr} 2 & 0 & 0 & -2 & 3 & 0 \\ 0 & 2 & 0 & 10 & -11 & 0 \\ 0 & 2 & -2 & 4 & -6 & 1 \end{array}\right]$$

Replace row two with the result.

$$\begin{array}{rrrrrrr} -R_2 \longrightarrow & 0 & -2 & 0 & -10 & 11 & 0 \\ R_3 \longrightarrow & 0 & 2 & -2 & 4 & -6 & 1 \\ \hline & 0 & 0 & -2 & -6 & 5 & 1 \end{array}$$

Add –1 times row two to row three.

$$\begin{array}{r} \\ \\ R_3 - R_2 \longrightarrow \end{array} \left[\begin{array}{rrr:rrr} 2 & 0 & 0 & -2 & 3 & 0 \\ 0 & 2 & 0 & 10 & -11 & 0 \\ 0 & 0 & -2 & -6 & 5 & 1 \end{array}\right]$$

Replace row three with the result.

$$\left[\begin{array}{ccc|ccc} 1 & 0 & 0 & -1 & \frac{3}{2} & 0 \\ 0 & 1 & 0 & 5 & -\frac{11}{2} & 0 \\ 0 & 0 & 1 & 3 & -\frac{5}{2} & -\frac{1}{2} \end{array}\right]$$

Divide rows one and two by 2. Divide row three by –2.

$$C^{-1} = \begin{bmatrix} -1 & \frac{3}{2} & 0 \\ 5 & -\frac{11}{2} & 0 \\ 3 & -\frac{5}{2} & -\frac{1}{2} \end{bmatrix}$$

The inverse has been found.

Practice Exercises

1. For the following matrix, **A**, find the values of elements

   (a) a_{12}
   (b) a_{23}
   (c) a_{32}

$$\mathbf{A} = \begin{bmatrix} 1 & 2 & -3 & 5 \\ 3 & 1 & -2 & 0 \\ 8 & 2 & 4 & -6 \end{bmatrix}$$

2. Given the following matrices:

$$\mathbf{A} = \begin{bmatrix} 4 & 2 \\ 0 & -2 \end{bmatrix} \qquad \mathbf{B} = \begin{bmatrix} 3 & -4 \\ 8 & 5 \end{bmatrix}$$

$$\mathbf{C} = \begin{bmatrix} 7 \\ 0 \\ 3 \end{bmatrix} \qquad \mathbf{D} = \begin{bmatrix} 5 \\ -1 \\ 9 \end{bmatrix}$$

   Find:
   (a) 3**A**

   (b) **6A + B**

   (c) 2**A** – 4**B**

   (d) **C – D**

   (e) 5**C** + 3**D**

3. Perform the following matrix multiplications.

   (a) $\begin{bmatrix} 4 \\ -5 \\ 8 \end{bmatrix} \cdot \begin{bmatrix} 6 & -3 & 1 \end{bmatrix}$

   (b) $\begin{bmatrix} 5 \\ 3 \end{bmatrix} \cdot \begin{bmatrix} 0 & -3 \end{bmatrix}$

   (c) $\begin{bmatrix} -1 & 1 \\ 0 & 2 \end{bmatrix} \cdot \begin{bmatrix} 2 & 1 & -8 \\ 0 & 3 & 1 \end{bmatrix}$

4. Find the determinant of the following matrices.

   (a) $\mathbf{A} = \begin{bmatrix} 3 & 4 \\ 0 & -1 \end{bmatrix}$

   (b) $\mathbf{B} = \begin{bmatrix} 6 & 2 \\ 1 & 5 \end{bmatrix}$

   (c) $\mathbf{C} = \begin{bmatrix} 3 & 2 & 6 \\ 0 & -1 & 2 \\ 1 & 5 & 2 \end{bmatrix}$

   (d) $\mathbf{D} = \begin{bmatrix} 2 & -4 & 0 \\ 4 & -1 & 6 \\ 0 & 2 & -3 \end{bmatrix}$

5. Find the inverse of the following matrices, if possible.

   (a) $A = \begin{bmatrix} 2 & 2 \\ -1 & 3 \end{bmatrix}$

   (b) $B = \begin{bmatrix} 1 & 0 & 1 \\ 0 & -1 & 2 \\ 2 & 1 & -1 \end{bmatrix}$

Answers

1. (a) 2

   (b) –2

   (c) 2

2. (a) $\begin{bmatrix} 12 & 6 \\ 0 & -6 \end{bmatrix}$

   (b) $\begin{bmatrix} 27 & 8 \\ 8 & -7 \end{bmatrix}$

   (c) $\begin{bmatrix} -4 & 20 \\ -32 & -24 \end{bmatrix}$

   (d) $\begin{bmatrix} 2 \\ 1 \\ -6 \end{bmatrix}$

   (e) $\begin{bmatrix} 40 \\ -3 \\ 42 \end{bmatrix}$

3. (a) $\begin{bmatrix} 24 & -12 & 4 \\ -30 & 15 & -5 \\ 48 & -24 & 8 \end{bmatrix}$

   (b) $\begin{bmatrix} 0 & -15 \\ 0 & -9 \end{bmatrix}$

   (c) $\begin{bmatrix} -2 & 2 & 9 \\ 0 & 6 & 2 \end{bmatrix}$

4. (a) –3

   (b) 28

   (c) –26

   (d) –66

5. (a) $\begin{bmatrix} \frac{3}{8} & -\frac{1}{4} \\ \frac{1}{8} & \frac{1}{4} \end{bmatrix}$

   (b) $\begin{bmatrix} -1 & 1 & 1 \\ 0 & -3 & -2 \\ 2 & -1 & -1 \end{bmatrix}$

3

Linear Programming

This chapter introduces the concepts of linear programmming. Among the topics covered will be translation of linear programming problems and application of rules. Other areas covered in this chapter are standard and non-standard systems, the simplex method, maximization, minimization, duality and slack variables.

3.1 DEFINITIONS AND CONCEPTS

Most problems that rely on solutions through linear programming are, by nature, word problems. The text of the problem sets the conditions from which a system of inequalities is built. The text also levies restrictions or "constraints" upon the system. The following definitions apply to that section of mathematics called **linear programming**.

A linear programming problem in two variables is defined by a linear expression, $w = ax + by$, called the objective function, subject to conditions expressed as linear inequalities, called constraints. The goal of "solving" a linear programming problem is to find the maximum or minimum value of w that satisfies the constraints.

> A **maximum** in a linear programming problem is the greatest w that can be obtained for $w = ax + by$ subject to the constraints on the system. Similarily, the **minimum** is the smallest w-value. The maximum or minimum occurs for particular values of (x, y) in the solution space.

We will now use an example to demonstrate how these definitions are employed. Let us suppose that a computer manufacturer can produce a combination of desktop computers and portable computers in a given week. Both types of computers use a certain type of "mother board," so the company can make no more than 500 computers a week. The desktop computers sell for \$1,500 each, while the laptop (portable) sells for \$2,000 each. The manufacturer has a weekly budget of \$400,000 from which it can make desktop computers at \$1,000 each and laptops at \$1,400 each. How many of each type of computer should the manufacturer make to maximize profit?

This problem has two variables: x = number of desktop computers and y = number of laptop computers. These are the two variables in the **objective function**. In order to maximize profit we need to find out how much profit is made on each type of computer. The desktop computer sells for \$1,500 and costs \$1,000 to make. The profit on each desktop computer is \$1,500 – \$1,000 = \$500. The laptop sells for \$2,000 and costs \$1,400 to make, so it gives a profit of \$600. Therefore, each desktop brings in \$500 and each laptop brings in \$600. The profit, P, is found by solving the complete **objective function**, $P = \$500x + \$600y$.

The **constraints** derive from the limits on money and parts available per week.

The computers each use one motherboard, so the total number of computers made must be less than or equal to the 500 motherboards available. This first constraint is $x + y \leq 500$. The total manufacturing cost must be less than or equal to \$400,000 per week. Thus, the second constraint is $\$1{,}000x + \$1{,}400y \leq 400{,}000$, since *each* desktop costs \$1,000 to produce and *each* laptop costs \$1,400 to produce. The final constraints stem from the seemingly obvious fact that the number of computers manufactured must be more than or at least equal to 0.

The linear equations and inequalities arising from this problem can be summarized as:

Maximize: $P = 500x + 600y$ (Objective Function)

Subject to: $1000x + 1400y \leq 400{,}000$

$x + y \leq 500$ (Constraints)

$x \geq 0, y \geq 0$

EXAMPLE 3.1

Find the objective function and the constraints for the following linear programming problems.

(a) A small furniture manufacturing company makes wooden chairs and wooden tables. Each table uses two sheets of lumber and each chair uses one sheet. There is enough storage for 200 sheets of lumber. Each table requires 6 hours labor and each chair 5 hours. There are 450 hours of labor available per week. If the profit for each table is $100 and for each chair is $40, how many chairs and tables should be made in order to maximize the profit for each week?

(b) A farmer has 200 acres in which to plant corn or soybeans. Each acre of corn brings a profit of $300 and uses 100 pounds of fertilizer. Each acre of soybeans brings a profit of $150 and uses 60 pounds of fertilizer. There is enough cash reserve to buy 10,000 pounds of fertilizer. It takes 5 hours per acre to plant and cultivate the corn and 8 hours per acre to plant and cultivate the beans. There are 750 hours available for planting and cultivating. How many acres of corn and soybeans should be planted in order to maximize the farmer's profit?

Steps Involved in Translating to Linear Functions

Before we begin the translation of the problem text to linear equations and inequalities, we will summarize the steps that are common to many solutions.

1. Take **time** to read the problem thoroughly. These problems are often very lengthy.
2. Assign variables to unknown quantities.
3. Build a table to organize data, keeping like units (money, time, material) together.
4. Define the objective function (often, but not always, a profit equation) as a linear expression.
5. Express all constraints as linear inequalities.

SOLUTION 3.1

(a) x = number of chairs made/week
y = number of tables made/week

After the problem is read thoroughly, assign the variables x and y.

Construct a table to organize the data.

	chair	table	limit value
defined variable	x	y	–
profit (dollars)	40	100	P
time (hours)	5	6	450
material (sheets)	1	2	200

$P = 40x + 100y$

The top two rows of the table combine to form the **objective function**.

$5x + 6y \le 450$
$1x + 2y \le 200$

The last two rows form the constraint inequalities.

$x \ge 0 \quad y \ge 0$

Of course, x and y must be equal to or greater than zero.

maximize: $P = 40x + 100y$
subject to: $5x + 6y \le 450$
$x + 2y \le 200$
$x \ge 0\ ;\ y \ge 0$

The translation is now complete.

(b) x = number of acres of corn
y = number of acres of soybeans

After the problem is read thoroughly, assign the variables x and y.

Construct a table to organize the data.

	corn	soybeans	limit value
defined variable	x	y	–
profit (dollars)	300	150	P
time (hours)	5	8	750
material (pounds)	100	60	10,000
land (acres)	1	1	200

$P = 300x + 150y$

The top two rows of the table combine to form the **objective function**.

$5x + 8y \le 750$
$100x + 60y \le 10000$
$x + y \le 200$

The last three rows form the constraint inequalities.

$x \ge 0 \quad y \ge 0$

Of course, x and y must be equal to or greater than zero.

maximize: $P = 300x + 150y$

The translation is now complete.

subject to: $5x + 8y \leq 750$
$100x + 60y \leq 10000$
$x + y \leq 200$
$x \geq 0;\ y \geq 0$

3.2 GRAPHICAL SOLUTIONS OF LINEAR PROGRAMMING PROBLEMS

Small linear programming problems lend themselves nicely to solutions by graphing the constraint linear inequalities and matching the objective function. The linear inequalities can (1) enclose or **bound** an area on a graph or (2) leave the area on the graph **unbounded**.

> The bounded area defined by constraints is the set of points in the **region of feasible solutions**.

> **Corner points** are the vertices of the boundaries.

> It is possible to show that optimal solutions always occur at vertices of the region of feasible solutions so the **optimal solution** is the corner point that yields the optimal value for the objective function.

The procedure for graphical solutions of linear programming problems is outlined below.

1. Use the **constraints** (linear inequalities) to define a region of feasible solutions.
2. Determine the **corner points** of the region.
3. Evaluate the **objective function** at the coordinates of the **corner points**. The largest value obtained is the maximum. The smallest value obtained is the minimum.

The solution process can be demonstrated using the following data:

Maximize:	$40x + 30y = P$	Objective function
Subject to:	$3x + 2y \le 12$	
	$\frac{1}{2}x + y \le 4$	Constraints
	$x \ge 0;\ y \ge 0$	

(Refer to chapter 1 for information on graphing inequalities).
Each of the constraints are graphed in Figure 3.1.

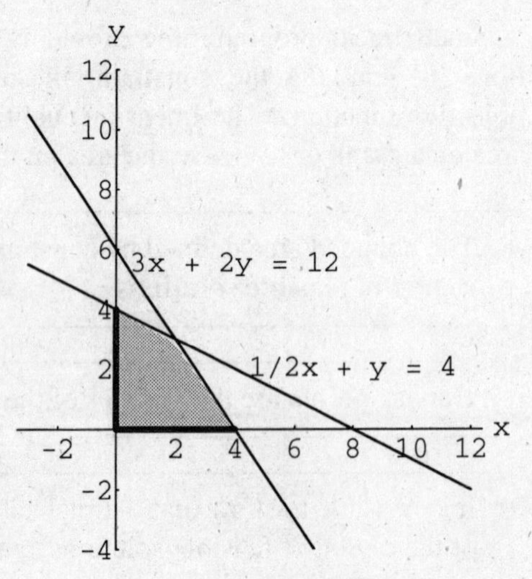

Figure 3.1

The four constraints enclose a region of feasible solutions. The corner points are labeled in Figure 3.2.

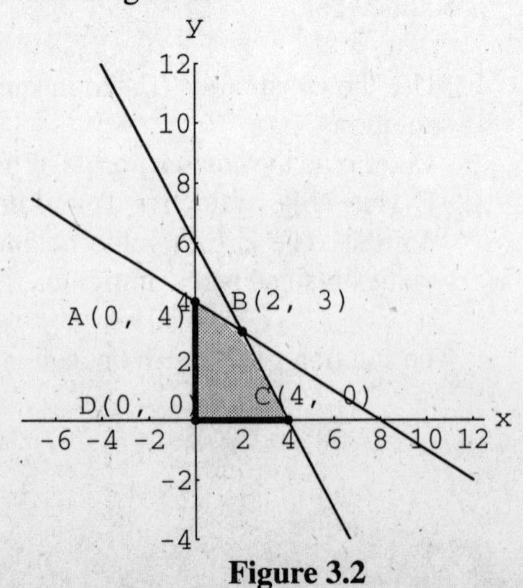

Figure 3.2

Point A is the intersection of the line $(1/2)x + y = 4$ with the line $x = 0$ (the y-axis). Point B is the intersection of the lines $3x + 2y = 12$ and $(1/2)x + y = 4$. We solved the two equations simultaneously to determine that (2, 3) is the solution set. Much like point A, point C is the intersection of $3x + 2y = 12$ with $y = 0$ (the x-axis). Point D is the intersection of the two axes at the origin. The origin will be a corner for any linear programming problem with $x \geq 0$, $y \geq 0$.

We now evaluate the objective function at the given corner points.

point (x, y)
A (0, 4)
B (2, 3)
C (4, 0)
D (0, 0)

The objective function $(40x + 30y = P)$ evaluated at each of the corner point is as follows:

at A: $40(0) + 30(4) = P$
$120 = P$

at B: $40(2) + 30(3) = P$
$80 + 90 = P$
$170 = P$

at C: $40(4) + 30(0) = P$
$160 = P$

at D: $40(0) + 30(0) = P$
$0 = P$

The maximum value for P is 170. Thus, the objective function $40x + 30y = 170$ intersects the set of feasible solutions at the best corner point ($x = 2$, $y = 3$).

Figure 3.3 shows $40x + 30y = 170$ and the set of feasible solutions.

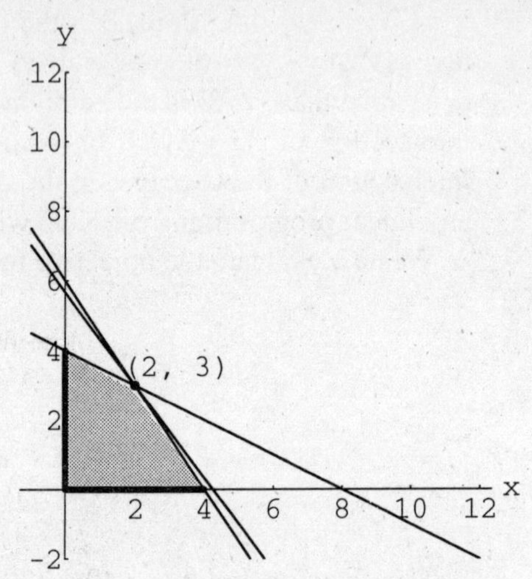

Figure 3.3

Note that the graph of the line $40x + 30y = 170$ intersects the region of feasible solutions at the corner point (2, 3). This figure summarizes the graphical solution of the linear programming problem.

EXAMPLE 3.2

Find the solution of the following linear programming problems using graphical methods.

(a) Maximize: $20x + 15y = P$

Subject to: $4x + y \geq 80$
$8x + 5y \leq 400$
$x \geq 10,\ y \geq 10$

(b) Minimize: $5x + 8y = P$

Subject to: $3x - 2y \leq 18$
$x + y \leq 4$
$x \geq 2,\ y \geq 0$

SOLUTION 3.2

(a) $\left.\begin{array}{l} 4x + y \geq 80 \\ 8x + 5y \leq 400 \\ x \geq 10,\ y \geq 10 \end{array}\right\}$ Constraints — List the constraints.

Graph all of the constraints in the first quadrant. To graph each constraint, treat it as if it were an equality to get the boundary line, *then* look at the inequality sign to determine which side of the line to shade.

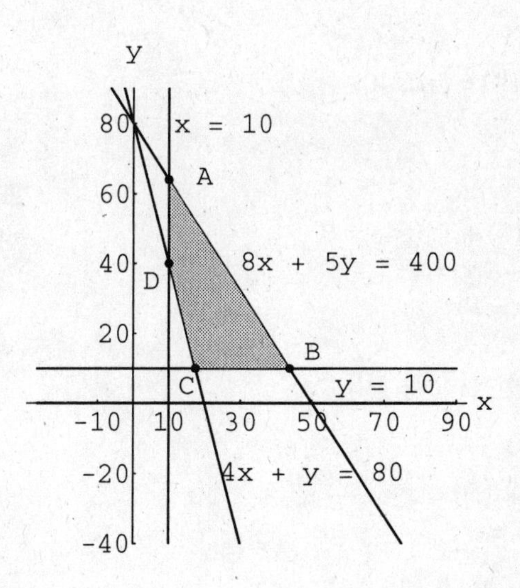

The four constraints enclose the region of feasible solution, F. The four corner points are A, B, C, and D.

The lines $8x + 5y = 400$ and $4x + y = 80$ do not intersect in the region of feasible solution. However, both of these lines do intersect $x = 10$ and $y = 10$. Simultaneous solutions for each pair of lines gives us the corner solutions.

$$\begin{aligned} 8x + 5y &= 400 \\ x &= 10 \end{aligned} \qquad \begin{aligned} 8x + 5y &= 400 \\ \underline{-8x \quad} &\underline{= -80} \\ 5y &= 320 \\ y &= 64 \end{aligned}$$

Point A = (10, 64)

$$\begin{aligned} 8x + 5y &= 400 \\ y &= 10 \end{aligned} \qquad \begin{aligned} 8x + 5y &= 400 \\ \underline{-5y} &\underline{= -50} \\ 8x \quad &= 350 \\ x &= 43.75 \end{aligned}$$

Point B = (43.75, 10)

$$\begin{array}{ll} 4x + y = 80 & 4x + y = 80 \\ y = 10 & \underline{\quad -y = -10} \\ & 4x \quad = 70 \\ & x = 17.5 \end{array}$$

Point C = (17.5, 10)

$$\begin{array}{ll} 4x + y = 80 & 4x + y = 80 \\ x \quad = 10 & \underline{-4x \quad = -40} \\ & y = 40 \end{array}$$

Point D = (10, 40)

The objective function, $20x + 15y = P$, is evaluated at each corner point.

For A: (10, 64)

$20\,(10) + 15(64) = P$
$200 + 960 = P$
$1160 = P$

For B: (43.75, 10)

$20(43.75) + 15\,(10) = P$
$875 + 150 = P$
$1025 = P$

For C: (17.5, 10)

$20\,(17.5) + 15\,(10) = P$
$350 + 150 = P$
$500 = P$

For D: (10, 40)

$20(10) + 40(10) = P$
$200 + 400 = P$
$600 = P$

The maximum value occurs at corner point, A. The objective function evaluated at point A gives the optimal value for $P = 1160$.

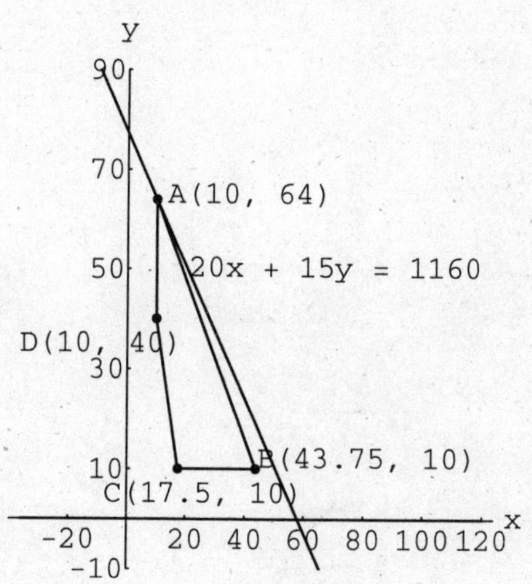

(b) $3x - 2y \leq 18$
$x + y \geq 4$ Constraints
$x \geq 2,\ y \geq 0$

List the constraints.

Graph them all in the first quadrant of the coordinate axis system. Treat each constraint as if it were an equality.

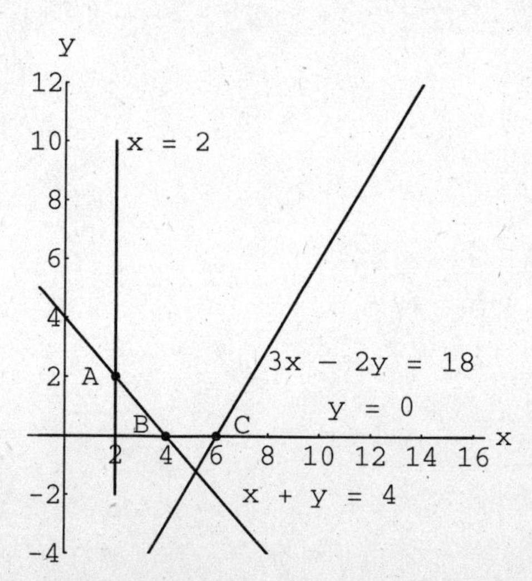

The four constraints do define, and completely enclose the region of feasible solution, F. There are three corner points are A, B, and C. Consequently the objective function cannot have a maximum but it can have a minimum.

For A:

$x + y = 4$

$x \quad = 2$

$$\begin{array}{r} x + y = 4 \\ \underline{-x \quad = -2} \\ y = 2 \end{array}$$

A = (2, 2)

Point A is the intersection of $x = 2$ and $x + y = 4$.

For B:

$x + y = 4$

$y = 0$

$$\begin{array}{r} x + y = 4 \\ \underline{-y = 0} \\ x \quad = 4 \end{array}$$

B= (4, 0)

Point B is the intersection of of $y = 0$ with $x + y = 4$.

For C:

$3x - 2y = 18$

$y = 0$

$$\begin{array}{r} 3x - 2y = 18 \\ \underline{2y = 0} \\ 3x \quad = 18 \\ x \quad = 6 \end{array}$$

Point C is the intersection of $y = 0$ with $3x - 2y = 18$.

C = (6, 0)

For A:

$5(2) + 8(2) = P$

$10 + 16 = P$

$26 = P$

The objective function, $5x + 8y = P$, is evaluated at each corner point.

For B:

$5(4) + 8(0) = P$

$20 = P$

For C:

$5 (6) + 8 (0) = P$

$30 = P$

$5x + 8y = 20$

The minimum occurs at corner point, B, (4,0). The

objective function evaluated at point B gives the optimal value for $P = 20$.

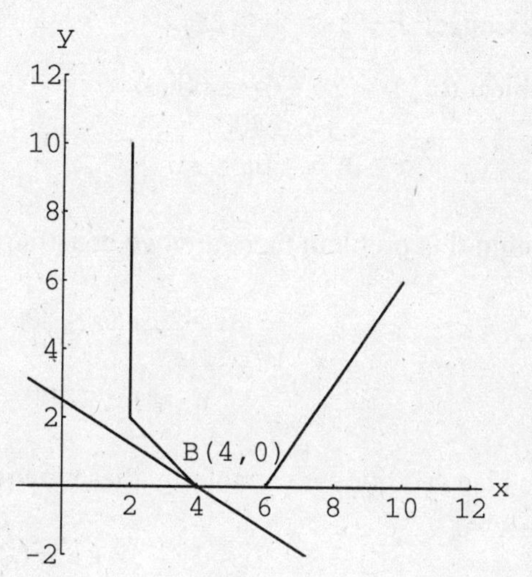

The line, $5x + 8y = 20$, intersects the corner point B on the region of feasible solution.

3.3 SLACK VARIABLES AND BASIC SOLUTIONS

While the previous section covered the graphical solution of linear programming problems, the next two sections will focus on other computational algorithms for solving linear programming problems. The solution process will focus on the **standard maximum problem** (SMP). The definition of SMP follows:

1. The variables x, y, and z are constrained to be non-negative.
2. All other constraints are of the form, $ax + by + cz + \ldots \leq P$ (where P is some positive constant).
3. The objective function is a linear function that must be **maximized**.

Slack Variables

A set of linear equations for a SMP might be as follows.

Maximize: $P = 3x + 5y - 2z$

Subject to: $3x - 2y + 6z \leq 6000$
$x - y + z \leq 400$
$x \geq 0;\ y \geq 0;\ z \geq 0$

Within this problem there are two constraints that involve several variables.

$$3x - 2y + 6z \leq 6000$$

and

$$x - y + z \leq 400$$

We place additional variables in these **inequalities** to convert them to **equations**.

$$3x - 2y + 6z + u = 6000$$

$$x - y + z + v = 400$$

The two added variables, u and v, are called **slack variables**. The other variables that were originally part of the inequalities are called the **decision variables.**

The use of a computational solution with slack variables is especially valuable in systems with **more than two** decision variables. Graphical solutions are more difficult when there are three or more decision variables.

Basic Solutions

Slack variables can be used to find **basic solutions**, which are nonnegative solutions to a linear programming problem. These solutions are those that occur at corners of the feasible solution set. In the following system, we will use slack variables to help us determine the corners of the feasible solution set.

Maximize: $P = 2x + 6y$

Subject to: $x + y \leq 6$
$x \leq 4$
$x + 2y \leq 10$
$x \geq 0, y \geq 0$

This is an SMP problem with three constraint inequalities:

$$x + y \leq 6$$
$$x \leq 4$$
$$x + 2y \leq 10$$

A slack variable is introduced for each inequality so that there are three equations formed from the constraints.

$$x + u = 4$$
$$x + y + v = 6$$
$$x + 2y + w = 10$$

If this is reordered with blank entries for the missing variables, we can begin to build the equations for an augmented matrix.

$$\begin{aligned} u \quad\quad\; + x \quad\;\; &= 4 \\ v \quad + x + y \;\; &= 6 \\ w + x + 2y &= 10 \end{aligned}$$

In the augmented matrix form, these equations should look like the following

$$\left[\begin{array}{ccccc|c} 1 & 0 & 0 & 1 & 0 & 4 \\ 0 & 1 & 0 & 1 & 1 & 6 \\ 0 & 0 & 1 & 1 & 2 & 10 \end{array}\right]$$

Note that this looks much like the completed matrix operation of chapter 2. The SMP has a **basic solution** when all decision variables equal zero. A corner point is (0, 0).

The above matrix gives the following **basic solution** if x and y are set to zero.

$$u = 4, v = 6, w = 10, x = 0, y = 0$$

The original set of equations gives us one basic solution. The equations

$$\begin{aligned} u \quad\quad\; + x \quad\;\; &= 4 \\ v \quad + x + y \;\; &= 6 \\ w + x + 2y &= 10 \end{aligned}$$

can be turned to another basic solution when x and u exchange places. This can be done by rewriting the first equation form and replacing the x in the other equations by $4 - u$.

$$\begin{aligned} x \quad\quad\; + u \quad\;\; &= 4 \\ v \quad - u + y \;\; &= 2 \\ w - u + 2y &= 6 \end{aligned}$$

When $u = 0$ and $y = 0$, $x = 4$, $v = 2$, $w = 6$
A second corner point is (4, 0).

The next **basic solution** (and corner point) is found by row operations that eliminate $2y$ from row 3. The result is

$$x \quad + u \qquad = 4$$
$$y \quad - u + v \ = 2$$
$$w + u + 2y = 2$$

with $u = 0$ and $v = 0$, then $x = 4$, $y = 2$, and $w = 2$ produces another corner point at (4, 2).

A third **basic solution** occurs when x, y, and u are isolated using row operations

$$u - \frac{3}{2}v - \frac{1}{2}w = 0$$
$$x \quad + v \qquad = 2$$
$$y - v + w = 4$$

When $v = 0$ and $w = 0$, a third basic solution is $u = 0$, $x = z$, $y = 4$. Now, (2, 4) is a third corner point.

Other row operations produce the corner points (4, 2) and (0, 5). These corner points can be seen in the graph of Figure 3.4.

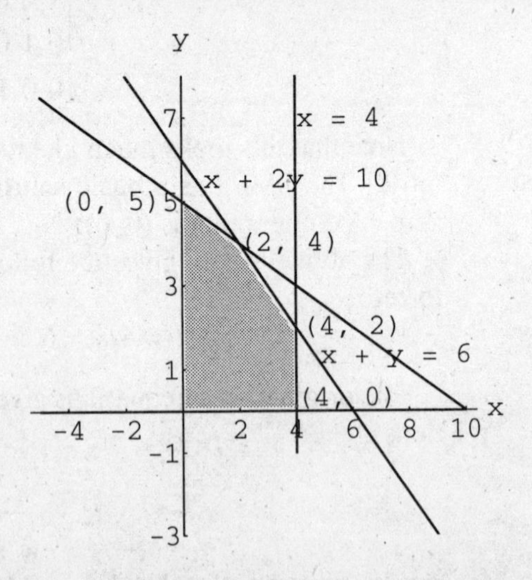

Figure 3.4

As you can see, the sequential process of seeking basic solution after basic solution (corner points) is very lengthy. The next section will seek a maximal solution in a more efficient manner.

3.4 MAXIMAL SOLUTIONS AND THE SIMPLEX TABLEAU

The simplex method operates through an array of numbers called a **tableau**. The method uses the tableau to determine each basic solution (corner point) in turn until the maximal solution is found. The development of an initial tableau is demonstrated by the following SMP.

$$P = 2x - 7y + 3z$$
$$x \geq 0 \quad y \geq 0 \quad z \geq 0$$
$$x + y - 3z \leq 100$$
$$2x + 5y + 4z \leq 150$$

The constraints involving the combination of decision variables are entered into the tableau first. The slack variables are introduced here.

$$\begin{array}{l} \\ \text{constraint} \\ \text{constraint} \end{array} \begin{array}{c} \begin{array}{cccccc} x & y & z & u & v & \text{solution} \end{array} \\ \left[\begin{array}{ccccc|c} 1 & 1 & -3 & 1 & 0 & 100 \\ 2 & 5 & 4 & 0 & 1 & 150 \end{array}\right] \end{array}$$

The tableau is not complete until the coefficients of the objective function are appended in a third row. The negatives of the objective function coefficients are placed in the third row. Initially, the objective function slack variables are set equal to 0 (zero).

$$\begin{array}{l} \\ \text{constraints} \\ \\ \text{objective function} \end{array} \begin{array}{c} \begin{array}{ccccc} x & y & z & u & v \end{array} \\ \left[\begin{array}{ccccc|c} 1 & 1 & -3 & 1 & 0 & 100 \\ 2 & 5 & 4 & 0 & 1 & 150 \\ -2 & 7 & -3 & 0 & 0 & 0 \end{array}\right] \end{array}$$

Above is the simplex tableau for this SMP.

Pivot Operation and Maximal Solution

The operations with the simplex tableau continue until all of the coefficients of the bottom row become positive.

We will need an example to explain how the "pivot" element is chosen and how the process proceeds.

Maximize: $P = 4x + 2y$

Subject to: $2x + 3y \leq 18$

$$3x + y \le 9$$
$$x \ge 0;\ y \ge 0$$

The initial tableau is set up as shown below.

$$\begin{array}{cccc|c} x & y & u & v & \text{sol.} \\ \hline 2 & 3 & 1 & 0 & 18 \\ 3 & 1 & 0 & 1 & 9 \\ -4 & -2 & 0 & 0 & 0 \end{array}$$

Each column is identified by possible "entry" variables. The rows will also be identified by "departing" variables (except the bottom row, which holds the objective function).

$$\begin{array}{c|cccc|c} & x & y & u & v & \text{sol.} \\ \hline u & 2 & 3 & 1 & 0 & 18 \\ v & 3 & 1 & 0 & 1 & 9 \\ P & -4 & -2 & 0 & 0 & 0 \end{array}$$

The first step in defining the "entry" variable is to select the column that has the **most negative** coefficient. In this example, x is the "entry" variable.

$$\begin{array}{c|cccc|c} & \downarrow x & y & u & v & \text{sol.} \\ \hline u & 2 & 3 & 1 & 0 & 18 \\ v & 3 & 1 & 0 & 1 & 9 \\ P & -4 & -2 & 0 & 0 & 0 \end{array}$$

The proper row is selected by dividing each solution element by the coefficient found in the x-column ("entry" variable column). The fraction $18/2$ in the u row is more than the fraction $9/3$ in the v-row. The row with the lesser value is our choice. The pivot element is $9/3$ (3).

$$\begin{array}{c|cccc|c} & \downarrow x & y & u & v & \text{sol.} \\ \hline u & 2 & 3 & 1 & 0 & 18 \\ \rightarrow v & 3 & 1 & 0 & 1 & 9 \\ P & -4 & -2 & 0 & 0 & 0 \end{array}$$

The x is swapped (or pivoted) for v around 3. This is accomplished by converting the element to a 1 in the pivot element location and zeros elsewhere. Use matrix row operations to complete this process. After the row operations are done we have

$$\begin{array}{c} \\ u \\ x \\ P \end{array}\begin{array}{c} \begin{array}{ccccc} x & y & u & v & \text{sol.} \end{array} \\ \begin{bmatrix} 0 & 7 & 3 & -2 & 36 \\ 1 & \frac{1}{3} & 0 & \frac{1}{3} & 3 \\ 0 & -1 & 0 & 1 & 9 \end{bmatrix} \end{array}$$

A -1 still exists in the bottom row, so we choose y as our departing variable. Again, dividing $36/7$ and 3/(1/3) we find that $36/7$ (or $5\ 1/7$) is **less** than 3/(1/3) (or 9). The pivot element is 7. It is circled below.

$$\begin{array}{c} \\ \longrightarrow u \\ x \\ P \end{array}\begin{array}{c} \begin{array}{ccccc} x & \overset{\downarrow}{y} & u & v & \text{sol.} \end{array} \\ \begin{bmatrix} 0 & \textcircled{7} & 3 & -2 & 36 \\ 1 & \frac{1}{3} & 0 & \frac{1}{3} & 3 \\ 0 & -1 & 0 & 1 & 9 \end{bmatrix} \end{array}$$

Again, ***y*** is the **entering** variable, and ***u*** is the **departing** variable. Row operations are used to place a 1 where the pivot element is. Zeroes are placed elsewhere.

$$\begin{array}{c} \\ y \\ x \\ P \end{array}\begin{array}{c} \begin{array}{ccccc} x & y & u & v & \text{sol.} \end{array} \\ \begin{bmatrix} 0 & 1 & \frac{3}{7} & -\frac{2}{7} & \frac{36}{7} \\ 1 & 0 & -\frac{1}{7} & \frac{3}{7} & \frac{9}{7} \\ 0 & 0 & \frac{3}{7} & \frac{5}{7} & \frac{99}{7} \end{bmatrix} \end{array}$$

The process stops when there are no more **negative** coefficients in the bottom row. This basic solution is the maximal one, with $x = 9/7$, $y = 36/7$, $u = v = 0$ and

$$P = 3x + 2y = 3\left(\frac{9}{7}\right) + 2\left(\frac{36}{7}\right) = \frac{27}{7} + \frac{72}{7} = \frac{99}{7}.$$

A graph of the system verifies that the corner point $\left(\frac{9}{7}, \frac{36}{7}\right)$ is maximal.

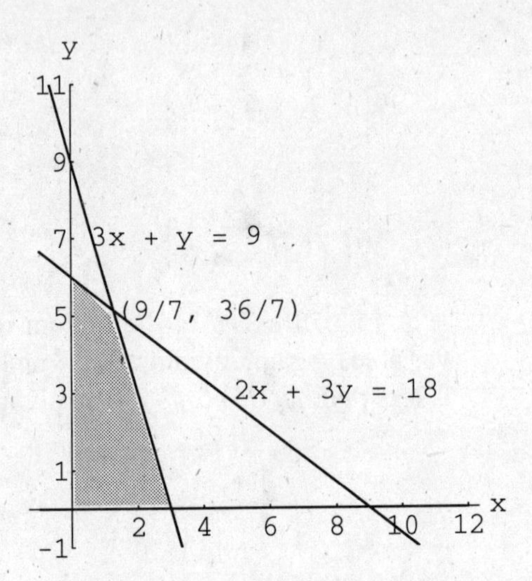

EXAMPLE 3.3

Use the maximality test and the simplex tableau to find the solution to the following linear programming problems (SMP form).

(a) Maximize: $8x + y = P$

Subject to: $x \geq 0, y \geq 0$
$x + y \leq 10$
$x + 3y \leq 15$

(b) Maximize: $x + 3y + 2z = P$

Subject to: $x \geq 0, y \geq 0, z \geq 0$
$x + 2y + 3z \leq 12$
$4x + 3y + 2z \leq 20$
$2x + y + z \leq 10$

SOLUTION 3.3

(a) Maximize: $8x + y = P$

Copy the pertinent information for this problem.

Subject to: $x \geq 0, y \geq 0$
$x + y \leq 10$
$x + 3y \leq 15$

$$\begin{array}{c} \\ u \\ v \\ P \end{array}\begin{array}{c} \begin{array}{ccccc} x & y & u & v & \text{sol.} \end{array} \\ \left[\begin{array}{cccc|c} 1 & 1 & 1 & 0 & 10 \\ 1 & 3 & 0 & 1 & 15 \\ -8 & -1 & 0 & 0 & 0 \end{array}\right] \end{array}$$

Form the simplex tableau by introducing the slack variables u and v. Also, enter the coefficients of the objective function as negative coefficients.

$$\begin{array}{c} \\ \rightarrow u \\ v \\ P \end{array}\begin{array}{c} \begin{array}{ccccc} \downarrow x & y & u & v & \text{sol.} \end{array} \\ \left[\begin{array}{cccc|c} \textcircled{1} & 1 & 1 & 0 & 10 \\ 1 & 3 & 0 & 1 & 15 \\ -8 & -1 & 0 & 0 & 0 \end{array}\right] \end{array}$$

The x-column has the most negative coefficient. The ratio $10/1$ is less than $15/1$.
Therefore, the circled element is the pivot element. x is the entering variable and u is the departing variable.

$$\begin{array}{c} \\ x \\ v \\ P \end{array}\begin{array}{c} \begin{array}{ccccc} x & y & u & v & \text{sol.} \end{array} \\ \left[\begin{array}{ccccc} 1 & 1 & 1 & 0 & 10 \\ 0 & 2 & -1 & 1 & 5 \\ 0 & 7 & 8 & 0 & 80 \end{array}\right] \end{array}$$

Matrix row operation produce a 1 in the pivot element and zeroes elsewhere in the column.

Since there are no more negative coeffients in the last row, the optimal solution is the basic solution $x = 10$, $y = 0$.

$$P = 8(x) + y = 8(10) + 1(0) = 80$$

The maximal solution is the last element in the bottom row (80).

(b) $x + 3y + 2z = P$

Copy the pertinent information for this problem.

$x \geq 0, y \geq 0, z \geq 0$
$x + 2y + 3z \leq 12$
$4x + 3y + 2z \leq 20$
$2x + y + z \leq 10$

$$\begin{array}{c} \\ u \\ v \\ w \\ P \end{array}\begin{array}{c} \begin{array}{ccccccc} x & y & z & u & v & w & \text{sol.} \end{array} \\ \left[\begin{array}{cccccc|c} 1 & 2 & 3 & 1 & 0 & 0 & 12 \\ 4 & 3 & 2 & 0 & 1 & 0 & 20 \\ 2 & 1 & 1 & 0 & 0 & 1 & 10 \\ -1 & -3 & -2 & 0 & 0 & 0 & 0 \end{array}\right] \end{array}$$

Form the simplex tableau by introducing the slack variables u, v, and w. Enter the negatives of the coefficients of the objective function in the last row.

$$\begin{array}{c} \\ u \\ v \\ w \\ P \end{array} \begin{array}{c} \begin{array}{ccccccc} x & y & z & u & v & w & \text{sol.} \end{array} \\ \left[\begin{array}{cccccc|c} 1 & \textcircled{2} & 3 & 1 & 0 & 0 & 12 \\ 4 & 3 & 2 & 0 & 1 & 0 & 20 \\ 2 & 1 & 1 & 0 & 0 & 1 & 10 \\ -1 & -3 & -2 & 0 & 0 & 0 & 0 \end{array}\right] \end{array}$$

The most negative coefficient is in the y-column. $\frac{12}{2} < \frac{20}{3} < \frac{10}{1}$ so the pivotal element is 2.

$$\begin{array}{c} y \\ v \\ w \\ P \end{array} \left[\begin{array}{cccccc|c} \frac{1}{2} & 1 & \frac{3}{2} & \frac{1}{2} & 0 & 0 & 6 \\ -5 & 0 & 5 & 3 & -2 & 0 & -4 \\ -3 & 0 & 1 & 1 & 0 & -2 & -8 \\ \frac{1}{2} & 0 & \frac{5}{2} & \frac{3}{2} & 0 & 0 & 18 \end{array}\right]$$

The pivot element is converted to 1 and the elements in that column converted to zeroes by row operations.

There are no more negative coefficients in the last row. Therefore, the basic solution at $x = 0$, $y = 6$, and $z = 0$ is the maximal solution.

$$P = 1x + 3y + 2z = 1(0) + 3(6) + 2(0) = 18$$

Practice Exercises

1. Find the objective function and the constraints for the following linear programming problems.

   (a) A construction company builds houses and condominiums. Each house brings a profit of $20,000 and each condominium brings a profit of $10,000. The cost of construction for a house is $50,000, while the condo minium costs $25,000. The time it takes to build a house is 15 weeks and to build a condominium is 12 weeks. The construction company has $650,000 to spend on construction for one year.

   (b) A farmer has room to plant 400 trees. The trees will be peach trees and pecan trees. Each peach tree brings a profit of $3000 and costs $150 to care for. Each pecan tree brings a profit of $2000 and cost $80 to care for. There is enough cash reserve to pay $30,000. It takes 50 hours per peach tree to care for and harvest peaches and it takes 35 hours per pecan tree to care for and harvest pecans. There are 7500 hours available for work. Maximize the farmer's profit.

2. Find the solution of the following linear programming problems using graphical methods.

   (a) Maximize: $100x + 85y = P$

   Subject to: $2x + y \leq 60$
   $x - y \leq 20$
   $x \geq 0\ ,\ y \geq 0$

   (b) Minimize: $25x + 13y = P$

   Subject to: $x + y \leq 8$
   $9x + y \leq 9$
   $x \geq 0,\ y \geq 0$

3. Use the maximality test and the simplex tableau to find the solution to the following linear programming problems (SMP form).

   (a) Maximize: $8x + y = P$

   Subject to: $x \geq 0,\ y \geq 0$
   $x + y \leq 10$
   $x + 3y \leq 15$

Answers

1. (a) $\$20{,}000x + \$10{,}000y = P$
$\$50{,}000x + \$25{,}000y \leq \$650{,}000$
$15x + 12y \leq 52$
$x \geq 0, y \geq 0$

(b) $\$3000x + \$2000y = P$
$x + y \leq 400$
$\$80x + \$150y \leq \$30000$
$50x + 35y \leq 7500$

2. (a) \$5700

(b) \$105.50

3. (a) \$80

4

Consumer Mathematics

There are several aspects of consumer mathematics that occur in our daily lives. These aspects invariably involve the saving or spending of money. The sections in this chapter will look at simple percents (such as you might see used in a retail store); compound interest (for banking operations); depreciation (for tax purposes).
Note: Some of the topics in this chapter have formulae that require the calculation of exponential expressions. These calculations can be done with a calculator or with a computer.

4.1 SIMPLE PERCENTS

Simple percent problems often arise for example when the price of a product is increased or discounted, or when a sales tax is applied. A 6% sales tax on an item would mean that an additional 6% of the cost of an item is added to the base price to get the total amount required.

> The **base price** of an item is the original cost of that item.

All percent operations involve either increasing or decreasing the base price.

Percents

Before we can use percents in the computation of simple interest, we must convert to the decimal number that is equal to the percent. This is done by dividing the given percent by 100% and thus finding the decimal equivalent.

Thus, 65% is converted to .65 in the following manner:

$$\frac{65\%}{100\%} = \frac{65}{100}$$

$$100\overline{)65}^{\,.65} \text{ gives } .65$$

The same result can be obtained in a manner that is considered to be a "shortcut." To use this method, simply drop the percent symbol and move the decimal to the **left** two places. The process for 65% follows:

65% ——> 65.% ——> .65. ——> .65

EXAMPLE 4.1

Convert the following percents to decimal numbers.

(a) 27%
(b) .53%
(c) 7.2%
(d) 147%

SOLUTION 4.1

(a) 27% ——> 27.% ——> .27. ——> .27

(b) .53% ——> .00.53 ——> .0053

(c) 7.2% ——> .07.2 ——> .072

(d) 147% ——> 147.% ——> 1.47. ——> 1.47

Note: Most calculators and spreadsheets will do this for you automatically.

Simple Percents

Simple percent problems require the multiplication of a converted decimal number by the base price of an item. In many problems involving simple percents, the word "of" represents multiplication.

The following problem involves simple percents.

EXAMPLE 4.2

Solve for the answer by converting from percent to decimal and multiplying.

(a) The price of oil increased by 15% per barrel over a period of ten years. What was the dollar increase, if the price was twenty dollars per barrel at the beginning of the ten-year period?

(b) A family bought a washing machine for $329. The sales tax is 6% in the state where they live. What was the tax, in dollars?

(c) A basketball cost $15.75 originally. However, the sporting goods store placed the basketball on sale at a 20% discount. What is the new price?

(d) Andy bought a disc player that cost $229.95. With a 7% sales tax, how much did Andy pay in total?

(e) The population of Nowhere, Nevada, grew 18% over the past decade. In the last census (a decade ago) the population was 250. What is the new population?

SOLUTION 4.2

(a) 15%——>15.0%——>15. ——>.15	Convert the percent to the decimal equivalent.
($20)(.15) = 3.00	Multiply the decimal number times the **base price**. This gives the dollar amount increase.
3.00——> $3 increase in price per barrel of oil.	
(b) 6%——>6.%——>6. ——>.06	Convert the percent to the decimal equivalent.
350.00(.06) = 21.00	Multiply the decimal number times the **base price**. This gives the amount of sales tax.

21.00—> $21.00
The amount of tax on the washing machine.

The sales tax of the washing machine is found.

(c) 20%—>20.%—>.20. —>.20

Convert the percent to the decimal equivalent.

(15.75)(.20) = 3.15

Multiply the decimal number times the **base price**. This gives the amount of the discount.

3.15—> $3.15

$15.75 – $3.15 = **$12.60**

Subtract the discount from the base price to get the **discounted price**.

(d) 7%—>7.%—>.07. —>.07

Convert the percent to the decimal equivalent.

(229.95)(.07) = 16.10

Multiply the decimal number times the **base price**. This gives the amount of sales tax.

229.95 + 16.10 = 246.05

= $246.05

The tax is added to the base price for the total selling price.

(e) 18%—>18.%—>.18. —>.18

Convert the percent to the decimal equivalent.

250(.18) = 45
45 is the population increase.

Multiplying the decimal number times the base population gives the population increase.

250 + 45 = 295
people in the new census.

This is added to the base to get the new population.

4.2 COMPOUND INTEREST

When an amount of money is deposited in a savings institution it can earn **interest**. The money deposited is called the **principal**. The interest earned is added to the principal. This new sum is then used as the principal for the next time interest is earned. This is an example of **compound interest**. The interest is found over a fixed time period called the **conversion period**.

There are essentially two types of formulae for compound interest. One is compounding for a finite (countable) number of conversion periods.

Countable Compounding per Year

$$A = P\left(1 + \frac{r}{n}\right)^{nt}$$

The other is the result of continuous compounding.

Continuous Compounding

$$A = Pe^{rt}$$

In the previous formulae:

P is the principal (amount of money deposited initially).
r is the interest rate (usually given in percent, usually per year).
t is the time in years.
n is the number of conversion periods per year.
A is the amount in the account after compounding.
e is the constant that is called the **natural number** (approximately equal to 2.71828).

EXAMPLE 4.3

(a) An amount of $1,000 is placed in an account at 7% interest rate. If the investment is compounded quarterly (every three months), how much is in the account after 2 years?

(b) An amount of $7,000 is placed in an account at 7% interest rate. If the investment is compounded monthly, how much is in the account after 5 years?

(c) An amount of \$6,000 is placed in an account at 8% interest rate. If the investment is compounded continuously, how much is in the account after 10 years?

(d) An amount of \$10,000 is placed in an account at 6% interest rate. If the investment is compounded continuously, how much is in the account after 5 years?

SOLUTION 4.3

(a) This problem uses the formula

$$A = P\left(1+\frac{r}{n}\right)^{nt}$$ because of the countable (4/yr.) conversion periods.

Here $P = \$1000$, $r = 7\%$, $t = 2$ and $n = 4$ (because of four quarters per year).

$$A = 1000\left(1+\frac{0.07}{4}\right)^{4(2)}$$

Convert 7% to .07. Substitute the values in the equation.

$$A = 1000(1+.0175)^8$$

Simplify the numbers in the equation.

$$A = 1000(1.0175)^8$$

The next step is to find the value of (1.0175) to the 8th power.

Enter 1.0175, then $\boxed{x^y}$ and then enter 8. Press the equal sign or execute key and get the result.
The result is 1.14889.

This can be done by entering 1.0175 into a calculator and pressing the x^y key shown at the left, then entering 8 to get the answer.

$$A = 1000(1.14889) = \$1{,}148.89$$

The product is the amount in the account after 2 years.

(b) This problem uses the formula

$$A = P\left(1+\frac{r}{n}\right)^{nt}$$ because of the countable (12) conversion periods.

Here $P = \$7000$, $r = 7\%$, $t = 5$ and $n = 12$ (because of twelve months per year).

$A = 7000\left(1 + \dfrac{0.07}{12}\right)^{12(5)}$

Convert 7% to .07. Substitute the values in the equation.

$A = 7000(1 + .00583)^{60}$

Simplify the numbers in the equation.

$A = 7000(1.00583)^{60}$

The next step is to find the value of (1.00583) to the 8th power.

Enter 1.00583, then $\boxed{x^y}$ and enter 60. Press the equal sign or execute key and get the result.
The result is 1.418.

This can be done by entering 1.00583 into a calculator and pressing the x^y key shown at the left, then enter 60 to get the answer.

$A = 7000(1.418) = \$9923.36$

The product is the amount in the account after 5 years.

(c) This problem uses the formula $A = Pe^{rt}$ because of the continuous compounding.

Here $P = \$6000$, $r = 8\%$, $t = 10$.

$A = 6000e^{(0.08)(10)}$

Convert 8% to .08. Substitute the values in the equation.

$A = 6000e^{.8}$

Simplify the numbers in the equation.

$A = 6000e^{.8}$

The next step is to find the value of e to the .8th power.

Press $\boxed{e^x}$ and then .8 to get 2.2255.

Press the exponential key. (It may require a 2nd function key or a shift key.) Then enter .8.

$A = (6000)(2.2255) = \$13{,}353.25$

Multiply the principal by the exponential value.

(d) This problem uses the formula $A = Pe^{rt}$ because of the continuous compounding.

Here $P = \$10000$, $r = 6\%$, $t = 5$ years.

$A = 10000e^{(0.06)(5)}$

Convert 6% to .06. Substitute the values in the equation.

$A = 10000e^{.3}$

Simplify the numbers in the equation.

$A = 10000e^{.3}$

The next step is to find the value of e to the .3 power.

Press $\boxed{e^x}$ and then .3 to get 1.3498.

Press the exponential key. (It may require a 2nd function key or a shift key.) Then enter .3.

$A = (10000)(1.3498) = \$13{,}498.59$

Multiply the principal by the exponential value.

4.3 DEPRECIATION

Buildings, vehicles, office equipment, etc. are considered to be business assets. As time goes on, the useful life of these assets is gradually used up. An accounting procedure is used to distribute the original cost of the equipment over the time period equal to the useful life of the equipment. This procedure is known as **depreciation**. The amount of money subtracted in each year of the procedure is also called the **depreciation**.

The amount of depreciation for one year is subtracted from the value adjusted from the previous year. The subtraction is continued until the amount remaining is the salvage value, or the depreciated cost is zero.

Straight-Line Depreciation

The **straight-line** method subtracts the salvage cost from the initial cost and divides that difference by the useful life of the item. The initial cost minus the salvage value is the **depreciation base**. Thus:

$$\text{Annual Straight-Line Depreciation} = \frac{\text{Depreciation Base}}{\text{Estimated Life}}$$

The estimated life is given in years.

EXAMPLE 4.4

(a) A stamping machine has an estimated life of 7 years. It cost $45,000 initially and can be salvaged for $3,000 at the end of its useful life. What is the annual depreciation? Assume straight-line depreciation.

(b) An office building has an estimated life of forty years. Its construction cost was $1,250,000. It is estimated that the building can be sold for $250,000 after 40 years. What is the annual depreciation using the straight-line depreciation method?

(c) An office copier is expected to last 5 years. Its initial cost is $1,200. There is expected to be no salvage value. What is the annual depreciation using the straight-line method?

SOLUTION 4.4

(a) $45,000 – $3,000 = $42,000

The **depreciation base** = initial cost minus salvage cost.

$$\text{depreciation} = \frac{42000}{7} = 6000$$

The straight-line depreciation is the depreciation base divided by the useful life (in years).

annual depreciation = $6,000.

(b) $1,250,000 – $250,000 = $1,000,000

The **depreciation base** = initial cost minus salvage cost.

$$\text{depreciation} = \frac{1000000}{40} = 25000$$

The straight-line depreciation is the depreciation base divided by the useful life (in years).

annual depreciation = $25,000.

(c) $\$1200 - \$0 = \$1200$

The **depreciation base** = initial cost minus salvage cost.

$$\text{depreciation} = \frac{1200}{5} = 240$$

The straight-line depreciation is the depreciation base divided by the useful life (in years).

annual depreciation = \$240.

Sum of the Year's Digits

This type of depreciation is used when equipment tends to depreciate more quickly at the beginning of its service life. In this method, the sum of the whole number digits of each year of service life is added to form the denominator of a multiplying fraction. The numerator of the fraction used as the multiplier is the largest single digit signifying the number of years remaining in the depreciation. For a piece of equipment that lasts 6 years, the sum of the years is $1 + 2 + 3 + 4 + 5 + 6 = 21$. The multiplier for each of the six years is as follows:

year	fraction
1	6/21
2	5/21
3	4/21
4	3/21
5	2/21
6	1/21

These fractions are then multiplied by the **depreciation base.**

EXAMPLE 4.5

(a) A machine that produces screws has an initial value of \$25,000. It has an estimated life of 8 years. Using the sum-of-the-digits method, what is the depreciation for the 5th year? The machine has no salvage value.

(b) A diesel truck is expected to last 5 years. It has a cost of \$23,000 when new. Using the sum-of-the-digits method, find the amount of depreciation in the third year. The salvage value is estimated at \$5,000.

(c) A cash register at a hardware store may last 3 years. It has no salvage value when the depreciation is done. If the cash register cost \$3,000 when new, what is the depreciation in the first year? Use the sum-of-the-digits method.

SOLUTION 4.5

(a) depreciation base = \$25,000 – 0 = \$25,000	Find the depreciation base by subtracting the **salvage value** from the **initial cost**.
1 + 2 + 3 + 4 + 5 + 6 + 7 + 8 = 36 for 8 years.	Add a digit representing each of the total years to get the sum of the digits.
$\frac{4}{36}(25{,}000)$ \$2,777.78	The numerator is the number of years remaining (4). This fraction is multiplied by the **depreciation base** to get the depreciation for that given year (year 5).
(b) depreciation base = \$23,000 – \$5,000 = \$18,000	Find the depreciation base by subtracting the **salvage value** from the **initial cost**.
1 + 2 + 3 + 4 + 5 = 15 for 5 years.	Add a digit representing each of the total years to get the sum of the digits.
$\frac{3}{15}(18{,}000)$ \$3,600	The numerator is the number of years remaining (3). This fraction is multiplied by the **depreciation base** to get the deprecieation for that given year (year 3).
(c) depreciation base = \$3,000 – 0 = \$3,000	Find the depreciation base by subtracting the **salvage value** from the **initial cost**.
1 + 2 + 3 = 6 for 3 years.	Add a digit representing each of the total years to get the sum of the digits.

$\frac{3}{6}(3{,}000)$	The numerator is the number of years remaining (3). This fraction is multiplied by the **depreciation base** to get the depreciation for that
\$1,500	given year (year 1).

4.4 AMORTIZATION

Home purchases are often computed by using a mortgage, where payment of the mortgage is made to a mortgage holder. **Amortization** is the payment of a debt through scheduled installments. The buyer of real estate needs a payment schedule so that the principal of the real estate and the interest charged on that principal are both reduced.

The payment found through the amortization formula shown below is the sum of the simple interest payment plus the amount paid toward the principal. The simple interest is the interest rate times the principal remaining.

$$R = P\left(\frac{i(1+i)^n}{(1+i)^n - 1}\right)$$

Where P is the principal.
i is the interest *per payment period.*
n is the number of pay periods.
R is the *rent* per period.

EXAMPLE 4.6

Solve the following amortization problems.

(a) What is the quarterly payment on a \$20,000 loan at 8% for 15 years?

(b) What is the monthly payment on a \$5,000 loan at 18% for 5 years?

(c) What is the actual payment toward principal on the first payment for a \$10,000 loan at 6% to be paid semiannually for 10 years?

SOLUTION 4.6

(a) $8\% = .08$

Convert the percent to a decimal.

$$\frac{0.08}{4} = 0.02 = i$$

The value of i is the annual percentage divided by the number of payments made annually.

$n = 4$ times $15 = 60$

The value of n is the number of payments made annually times the number of years needed for total payment.

$$R = P\left(\frac{i(1+i)^n}{(1+i)^n - 1}\right)$$

The formula needed is the one for amortization. Substitute in the values.

$$R = 20000\left(\frac{0.02(1+0.02)^{60}}{(1+0.02)^{60} - 1}\right)$$

$$= 20000\left(\frac{0.02(1.02)^{60}}{(1.02)^{60} - 1}\right)$$

We use the calculator to find $(1.02)^{60}$. Press 1.02, then x^y, then 60. Then = or Exe.

$$= 20000\left(\frac{0.02(3.28)}{3.28 - 1}\right)$$

The value of the exponential expression is 3.28.

$$= 20000\left(\frac{0.0656}{2.28}\right)$$

$= 20000(.02877)$
$= \$575.62$

The final value is the amortization payment.

(b) $18\% = .018$

Convert the percent to a decimal.

$$\frac{0.018}{12} = 0.015 = i$$

The value of i is the annual percentage divided by the number of payments made annually.

$n = 12$ times $5 = 60$

The value of n is the number of payments made annually times the number of years needed for total payment.

$$R = P\left(\frac{i(1+i)^n}{(1+i)^n - 1}\right)$$

The formula needed is the one for amortization. Substitute in the values.

$$R = 5000\left(\frac{0.015\,(1+0.015)^{60}}{(1+0.015)^{60} - 1}\right)$$

$$= 5000\left(\frac{0.015\,(1.015)^{60}}{(1.015)^{60} - 1}\right)$$

We use the calculator to find $(1.015)^{60}$. Press 1.015, then x^y, then 60.

$$= 5000\left(\frac{0.015\,(2.443)}{2.443 - 1}\right)$$

The value of the exponential expression is 2.443.

$$= 5000\left(\frac{0.03665}{1.443}\right)$$

$$= 5000(.0254)$$

$$= \$126.98$$

The final value is the amortization payment.

(c) $6\% = .06$

Convert the percent to a decimal.

$$\frac{0.06}{2} = 0.03 = i$$

The value of i is the annual percentage divided by the number of payments made annually.

$n = 2$ times $10 = 20$

The value of n is the number of payments made annually times the number of years needed for total payment.

$$R = P\left(\frac{i(1+i)^n}{(1+i)^n - 1}\right)$$

The formula needed is the one for amortization. Substitute in the values.

$$R = 10000\left(\frac{0.03\,(1+0.03)^{20}}{(1+0.03)^{20}-1}\right)$$

$$= 10000\left(\frac{0.03\,(1.03)^{20}}{(1.03)^{20}-1}\right)$$

We use the calculator to find $(1.03)^{20}$. Press 1.03, then x^y, then 20.

$$= 10000\left(\frac{0.03\,(1.8061)}{1.8061-1}\right)$$

The value of the exponential expression is 1.8061.

$$= 10000\left(\frac{0.05418}{0.8061}\right)$$

$= 10000(.06721)$
$= \$672.10$

The final value is the amortization payment.

$10000(6\%) = 10000(.06)$
$= \$600$

Multiply the total cost by the percentage to get the simple interest.

$\$672.21 - \$600.00$

Subtract the simple interest from the monthly payment to get the amount paid toward principal in the first payment.

\$72.21 paid toward principal.

Practice Exercises

1. Convert the following percents to decimal numbers.

   (a) 35%
   (b) .83%
   (c) 5.9%
   (d) 345%

2. Solve for the answer by converting from percent to decimal and multiplying.

   (a) The price of a shirt is decreased by 25% . What was the dollar increase, if the price was $40 originally.

   (b) A woman bought a motorcycle that listed for $1,350. The sales tax is 5%. What was the total cost (including tax)?

   (c) A factory noticed a 10% increase in productivity. If 3,454 items had been produced in a certain time period, how many items are now produced in the same time period?

3. (a) An amount of $2,000 is placed in an account at 6% interest rate. If the investment is compounded quarterly (every 3 months), how much is in the account after 3 years?

   (b) An amount of $1,000 is placed in an account at 5% interest rate. If the investment is compounded monthly, how much is in the account after 5 years?

   (c) An amount of $10,000 is placed in an account at 8% interest rate. If the investment is compounded continuously, how much is in the account after 8 years?

4. (a) A dump truck has an estimated life of 9 years. It cost $45,000 initially and can be salvaged for $8,000 at the end of its useful life. What is the annual depreciation? Assume straight-line depreciation.

   (b) An condominium has an estimated life of sixty years. Its construction cost was $250,000. It is estimated that the building can be sold for $100,000 after 60 years. What is the annual depreciation using the straight-line depreciation method?

   (c) An riding is expected to last 3 years. Its initial cost is $1,500. There is expected to be no salvage value. What is the annual depreciation using the straight-line method?

5. (a) A rug loom has an initial value of $425,000. It has an estimated life of 18 years. Using the sum-of-the-digits method, what is the depreciation for the 11th year? The machine has a salvage value of $100,000.

   (b) A tow truck is expected to last 8 years. It has a cost of $43,000 when new. Using the sum-of-the-digits method, find the amount of depreciation in the 5th year. The salvage value is estimated at $5,000.

Answers

1. (a) .35

   (b) .0083

   (c) .059

   (d) 3.45

2. (a) $30

   (b) $1417.50

   (c) 3800

3. (a) $2391.24

   (b) $1283.36

   c) $18,964.81

4. (a) $6166.67

   (b) $2500

   (c) $500

5. (a) $1169.59

   (b) $4222.22

5

Probability

We hear the word "probability" used often in our everyday experiences. Statements such as "What's the probability of that happening?" or "The probability of rain today is sixty percent" are just two examples.

5.1 DEFINITIONS

Within the context of this chapter, **probability** is the measure of how likely an event is to occur. This measure is numerical.

The Numerical Aspects of Probability

The probability of something happening is expressed as a **fractional** or **decimal** value between 0 and 1. A **zero** (0) **probability** means that something **will not occur**. A **one** (1) **probability** means that something **will certainly happen**. Most events have probabilities somewhere between zero and one.

Establishing Probabilities

Establishing a numerical value for a probability can be done through:

(a) an educated estimate
(b) an empirical process
(c) a theoretical process

An **educated estimate** assigns a value to a probability based upon past experiences. You have probably encountered situations in which you are so familiar with the outcomes, you could almost "predict" the results.

The **empirical process** is an experimental one in which several tries have produced a history for the event. A baseball player with a batting average of .315 has shown that 31.5% of the time he goes to bat, he gets a hit. Many probabilities are found **experimentally** by actual duplication of a process.

In this chapter, we will focus on the **theoretical processes** of finding simple probabilities. To establish these probabilities, we must be able to look at all predictable and possible outcomes.

Sample Size and Sample Space

The complete collection of all possible outcomes of a particular event is called the **sample space.**

The **sample size** is the number of all possible outcomes in the sample space.

Determining Probabilities

The process whereby we establish a probability is then:

Step 1. Determine the **theoretical** nature of the event.
Step 2. List all possible outcomes (**sample space**).
Step 3. Count all of the possible outcomes (**sample size**).
Step 4. Use the sample size and sample space to find a probability.

Two procedures used to determine a sample space and find a sample size are (1) a **probability tree** and (2) the **Fundamental Counting Principle**. These procedures are discussed in the next two sections of the the chapter.

5.2 THE FUNDAMENTAL COUNTING PRINCIPLE

The Fundamental Counting Principle is stated as follows:

> If there are f ways to perform one task, g ways to do another task, and h ways to do a third task, then the total number of ways to perform all tasks is:
>
> $f \times g \times h$

The above rule can be adjusted so that the number of ways to perform any multi-step task can be counted.

Relating the Fundamental Counting Principle to the situation of tossing a coin three times, each coin toss can be considered as occurring in one of two possible ways (one way for "heads" and one way for "tails"). Therefore, for three tosses the number of possible outcomes is:

$$2 \text{ ways} \times 2 \text{ ways} \times 2 \text{ ways} = 8 \text{ outcomes}$$

which equals the sample size we found with the probability tree.

EXAMPLE 5.1

(a) A restaurant serves 3 types of salad, 4 types of sandwich, and 8 types of drink. How many different ways can someone order a salad and sandwich with a drink?

(b) A ski shop sells 4 types of skis, 5 brands of boots, and 3 brands of poles. How many different ways can a customer order skis, boots, and poles?

(c) A large ensemble is going to form quartets. If there are 4 horn players, 6 cellists, 3 violinists, and 7 people who play the oboe, how many ways can one quartet be formed? (Assume one instrument of each type is used in a quartet.)

(d) Thirty boys and 20 girls attend a dance. How many ways can a boy and a girl be paired to dance?

(e) Twelve people are members of a club. How many ways can a president, vice president, and treasurer be elected if all the members are available to run for the offices? No person can hold more than one office.

SOLUTION 5.1

(a) There are 3 ways to get a salad, 4 ways to order a sandwich, 8 ways to order a drink.

$$\text{Total ways} = 3 \text{ ways} \times 4 \text{ ways} \times 8 \text{ ways} = \mathbf{96 \text{ ways}}$$

(b) There are 4 ways to order skis, 5 ways to order boots, and 3 ways to order poles.

$$\text{Total ways} = 4 \text{ ways} \times 5 \text{ ways} \times 3 \text{ ways} = \mathbf{60 \text{ ways}}$$

(c) There are 4 ways to obtain a horn player, 6 ways to select a cellist, 3 ways to find a violinist, and 7 ways to find an oboist.

Total ways = 4 ways × 6 ways × 3 ways × 7 ways = **504 ways**

(d) There are 30 ways to pick a boy and 20 ways to pick a girl.

Total ways = 30 ways × 20 ways = **600 ways**

(e) Even though the election may be simultaneous for all 3 offices, it is useful to view the election of officers as sequential. There are 12 ways to elect someone to the first office. Let us assume that this office is the club president. Once the president is selected, 11 people are available for vice president (the second office). After the vice president is elected, 10 people remain to run for treasurer (the third office).

Total ways = 12 ways × 11 ways × 10 ways = **1320 ways**

5.3 COMBINATIONS

In order for us to compute probabilities, we must first find the number of ways something can happen.

Factorial Problems

Combinations can be used to find the number of ways some processes can happen. A discussion of combinations will begin after we have introduced the mathematical concept of a **factorial**.

A **factorial** is the multiplication of a whole number, n, by every **positive** whole number less than n. Thus:

$$n! = n \times (n-1) \times (n-2) \times (n-3) \times \ldots \times 1$$

where the symbol ! means factorial.

Some examples of factorials:
0! = 1 (by definition)

$1! = 1$

$2! = 2 \cdot 1 = 2$

$3! = 3 \cdot 2 \cdot 1 = 6$

$4! = 4 \cdot 3 \cdot 2 \cdot 1 = 24$

$5! = 5 \cdot 4 \cdot 3 \cdot 2 \cdot 1 = 120$

Rules for Combinations

Combinations count a selection from a collection of objects.

A **combination** is the number of ways to select ***r*** objects from a group of ***n*** objects when **order of selection is not important**. The formula for a **combination** is:

$$C_r^n = \frac{n!}{r!\,(n-r)!}$$

C_r^n means "the number of ways to choose ***r*** objects out of n objects available."

n is the total number of objects
r is the number of objects chosen
(*n* – *r*) is the number of objects left over
! means factorial

EXAMPLE 5.2

(a) A coach has 7 people trying for 5 positions on a basketball team. How many ways can the coach select the players, if order is not important?

(b) Eight people are running a race to qualify for a position on a team. If the first 3 finishers qualify, how many combinations are possible? (Order is not important.)

(c) Each student is to select 2 books from a reading list of 8. How many combinations are possible to select? (Order is not important.)

(d) A parking lot is filled with 12 cars. If an employee of a large firm is to select one car to drive on a sales trip, how many selections are possible?

SOLUTION 5.2

(a) A coach is selecting 5 players out of 7.

$$\text{Combination} = C^7{}_5 = \frac{7!}{5!\,(7-5)\,!} = \frac{7!}{5!2!}$$

$$= \frac{7\cdot 6\cdot 5\cdot 4\cdot 3\cdot 2\cdot 1}{5\cdot 4\cdot 3\cdot 2\cdot 1\cdot 2\cdot 1} = \frac{7\cdot 6}{2\cdot 1} = \frac{42}{2}$$

= 21 ways

(b) Three (3) ways are selected out of 8.

$$\text{Combination} = C^8{}_3 = \frac{8!}{3!\,(8-3)\,!} = \frac{8!}{3!5!}$$

$$= \frac{8\cdot 7\cdot 6\cdot 5\cdot 4\cdot 3\cdot 2\cdot 1}{3\cdot 2\cdot 1\cdot 5\cdot 4\cdot 3\cdot 2\cdot 1} = \frac{8\cdot 7\cdot 6}{3\cdot 2\cdot 1} = \frac{56}{1}$$

= 56 ways

(c) Two (2) books are selected out of eight (8).

$$\text{Combination} = C^8{}_2 = \frac{8!}{2!\,(8-2)\,!} = \frac{8!}{2!6!}$$

$$= \frac{8\cdot 7\cdot 6\cdot 5\cdot 4\cdot 3\cdot 2\cdot 1}{2\cdot 1\cdot 6\cdot 5\cdot 4\cdot 3\cdot 2\cdot 1} = \frac{56}{2}$$

= 28 ways

(d) There is one (1) car chosen from twelve (12).

$$C^{12}{}_1 = \frac{12!}{1!\,(12-1)\,!}$$

$$= \frac{12\cdot 11\cdot 10\cdot 9\cdot 8\cdot 7\cdot 6\cdot 5\cdot 4\cdot 3\cdot 2\cdot 1}{1\cdot 11\cdot 10\cdot 9\cdot 8\cdot 7\cdot 6\cdot 5\cdot 4\cdot 3\cdot 2\cdot 1} = \frac{12}{1}$$

= 12 ways

One would expect there to be 12 ways to select one of the cars, since there are 12 cars to choose from.

5.4 PERMUTATIONS

A **permutation** is the number of ways to select ***r*** objects from a group of ***n*** objects when **order of selection is important**. The formula for a **permutation** is written:

$$_{n}P_{r} = \frac{n!}{(n-r)!}$$

where ***n*** is the total number of objects
r is the number of objects chosen
(***n*** – ***r***) is the number of left-over objects

Again, the symbology involving $_{n}P_{r}$ is somewhat unusual. One must remember the use of **P** for **permutations**.

Races with "place" finishes are situations in which order is important. For example, if three horses; A, B, and C compete to see who finishes first or second, then the outcomes could be as follows.

Outcome #	1st Place Finisher	2nd Place Finisher
1	horse A	horse B
2	horse B	horse A
3	horse A	horse C
4	horse C	horse A
5	horse B	horse C
6	horse C	horse B

There are 6 different ways the outcomes can occur **if order makes a difference**. If order is not important, then combinations should be used, and the number of ways to finish reduces to 3. Thus, this serves as an example of the fact that the number of **permutations** is greater than, or at least equal to, the number of **combinations**.

EXAMPLE 5.3

(a) Seven dogs run a race for 1st, 2nd, and 3rd places. How many ways can the dogs finish 1st, 2nd, and 3rd?

(b) A club has 12 members, with all running for president, vice-president, and treasurer. How many ways can the election turn out?

(c) The 5 people with the highest grade averages receive a certificate with their name on it. If there are 20 people in the class, how many ways can the certificates be awarded? The certificates are different colors, so order is important.

(d) Ten counties in southern Virginia are taking a census. They want to see which 3 have the greatest populations. How many different possible outcomes are there?

SOLUTION 5.3

(a) Order is important in a race. We choose 3 out of 7.

$$_{7}P_{3} = \frac{7!}{(7-3)!} = \frac{7!}{4!} = \frac{7 \cdot 6 \cdot 5 \cdot 4 \cdot 3 \cdot 2 \cdot 1}{4 \cdot 3 \cdot 2 \cdot 1}$$

$$= 7 \cdot 6 \cdot 5 = 210 \text{ ways}$$

(b) Order is important in a club election. Three (3) positions out of 12 are chosen.

$$_{12}P_{3} = \frac{12!}{(12-3)!} = \frac{12!}{9!} = \frac{12 \cdot 11 \cdot 10 \cdot 9 \cdot 8 \cdot 7 \cdot 6 \cdot 5 \cdot 4 \cdot 3 \cdot 2 \cdot 1}{9 \cdot 8 \cdot 7 \cdot 6 \cdot 5 \cdot 4 \cdot 3 \cdot 2 \cdot 1}$$

$$= 12 \cdot 11 \cdot 10 = 1320 \text{ ways}$$

Refer to the Example 2e) under the Fundamental Counting Principle. Note that the same answer was obtained two different ways.

(c) There are 5 students out of 20 chosen. The Permuation formula is used because order is important.

$$_{20}P_{5} = \frac{20!}{(20-5)!} = \frac{20!}{15!} = \frac{20 \cdot 19 \cdot 18 \cdot 17 \cdot 16 \cdot 15!}{15!}$$

$$= 20 \cdot 19 \cdot 18 \cdot 17 \cdot 16 = 1{,}860{,}480 \text{ ways}$$

(d) The order is important for the 10 counties. We choose 3 out of 10.

$${}_{10}P_3 = \frac{10!}{(10-3)!} = \frac{10!}{7!} = \frac{10 \cdot 9 \cdot 8 \cdot 7!}{7!}$$

$$= 10 \cdot 9 \cdot 8 = 720 \text{ ways}$$

5.5 SIMPLE PROBABILITY

The previous sections covered various ways to count the number of ways events can occur in the sample space. As mentioned earlier, the sample size is the **number of possible outcomes** of an **event**.

> A **probability** is the ratio of the **outcomes favorable to the event** to the **total number of possible outcomes.**

For a **simple** event:

$$P(a) = \text{probability} = \frac{\text{number of outcomes of interest}}{\text{number of possible outcomes in event A}}$$

$P(a)$ is read "The probability of event A."

EXAMPLE 5.4

Find the probability of the following simple events.

(a) There are 5 girls and 3 boys in a group of 8 children. Find the probability of randomly selecting a girl.

(b) A box contains 3 red cubes and 12 yellow cubes. What is the probability of randomly selecting a red cube?

(c) A container has 1 red cube, 3 yellow cubes, and 6 blue cubes. What is the probability of drawing a yellow cube?

(d) If the U.S. Senate has 56 Democrats and 44 Republicans, what is the probability of randomly selecting a Republican's name from a list of all the Senator's names?

(e) A board contains evenly spaced and randomly mixed colored pegs. It has 36 red pegs, 18 blue pegs, and 15 white pegs. If one tosses a hoop at the board, what is the probability that the hoop lands on a blue peg?

(f) A bag contains 3 blue chips, 5 red chips, and 1 white chip. What is the probability of selecting a green chip?

SOLUTION 5.4

(a) There are 5 girls (outcome of interest) of the 8 children (total outcomes).

$$P(\text{a}) = P\text{ (of selecting a girl)} = \frac{\text{outcome of interest}}{\text{total outcome}} = \frac{5}{8}$$

(b) The outcome of interest is selecting a red cube. There are 3 red cubes. The total number of possible outcomes is the number of cubes (15).

$$P(\text{a}) = P\text{ (of selecting a red cube)} = \frac{\text{outcome of interest}}{\text{total possible outcomes}}$$

$$= \frac{3}{15} = \frac{1}{5}$$

(c) The number of outcomes of interest is 3 (selecting a yellow cube). The total number of outcomes is 1 red + 3 yellow + 6 blue = 10 cubes.

$$P(\text{a}) = P\text{ (of selecting a yellow cube)} = \frac{\text{outcome of interest}}{\text{total number of outcomes}}$$

$$= \frac{3}{10}$$

(d) The number of outcomes of interest is 44 (selecting a Republican name). There are 100 total names.

$$P(\text{a}) = P\text{ (of selecting a Republican's name)}$$

$$= \frac{\text{outcome of interest}}{\text{total number of outcomes}} = \frac{44}{100} = \frac{11}{25}$$

(e) There are 18 blue pegs (outcomes of interest) of a total of 36 + 18 + 15 = 69 pegs.

$$P(\text{a}) = P\text{ (of selecting a blue peg)} = \frac{\text{outcome of interest}}{\text{total outcomes}}$$

$$= \frac{18}{69} = \frac{6}{23}$$

(f) Since there are no green chips in the sample space, the number of outcomes of interest is zero. The total of possible outcomes is $3 + 5 + 1 = 9$.

$$P(\text{a}) = P\text{ (of selecting a green)} = \frac{0}{9} = 0$$

Rules involving probabilities that hold for all events:

1. If an event **cannot happen**, its probability is **zero**.

2. If an event must **happen**, its probability is **one**.

3. All event probabilities satisfy $0 \leq P(A) \leq 1$.

4. The probability of an event **not** happening is one minus the probability that an event **will** happen. For event A it is: $1 - P(\text{A}) = P(\text{A}')$ = P(**complement** of A). If $P(\text{A}) = 0.7$, the complement of event A, A′, has a probability of $P(\text{A}') = 1 - P(\text{A}) = 1 - 0.7 = 0.3$.

Note: Some texts use $P(\overline{\text{A}})$ instead of $P(\text{A}')$ as a symbol for complement.

5.6 COMPOUND EVENT PROBABILITIES

> A **compound event** consists of more than one simple event.

These compound events can combine simple events through the process of **addition** or **multiplication**, depending upon the nature of the compound event.

"Event" needs to be redefined in light of its use in this section.

> An **event** is something that can occur.

Addition Rules

Determining the probability of compound events sometimes requires special formulas such as the **General Addition Rule**. While the formula involves the addition of event probabilities, it is used when we want to find the probability that one event **or** another will happen. The key word is "**or**."

> **The General Addition Rule**: Given event A and event B, the probability of event A happening or event B happening is given by:
>
> $P(\text{A } \textbf{or} \text{ B}) = P(\text{A}) + P(\text{B}) - P(\text{A and B})$

This rule is paraphrased by stating that "The probability of event A **or** event B happening is equal to the sum of their individual probabilities minus the probability that they happen **simultaneously**."

Two events that can happen at the same time are **not mutually exclusive**. This concept has roots in the ideas of set theory presented in Chapter 3 of this text. When events (sets) intersect each other, they are not mutually exclusive. See Figure 5.1.

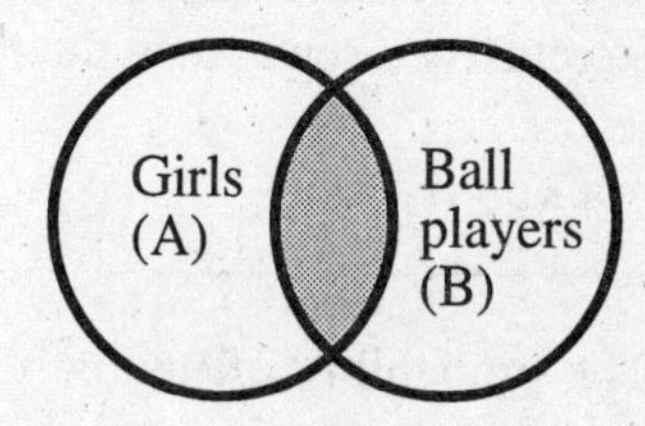

Figure 5.1 Intersecting Events Are Not Mutually Exclusive.

Here, it shows someone can be a girl and a ballplayer at the same time. Examples of events that are **not mutually exclusive** are as follows:

a bicyclist and someone over age seven

a teacher and someone who is married

an even number and a number above 5

Events that cannot happen at the same time are mutually exclusive events. These can be described by sets that are disjoint (see Chapter 3). Figure 5.2 shows mutually exclusive events.

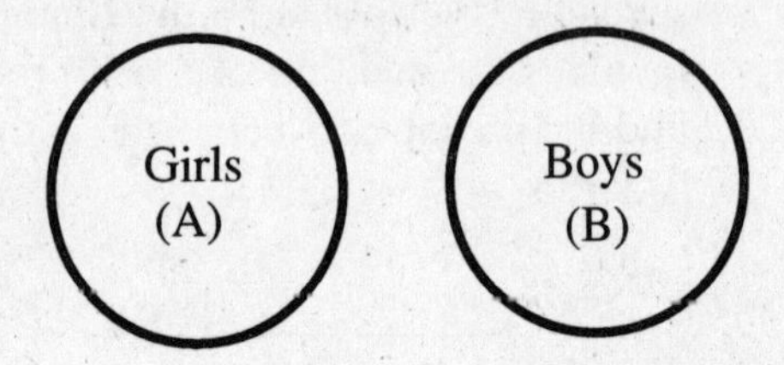

Figure 5.2 Mutually Exclusive Events Are Disjoint

These events are mutually exclusive because they can not possibly occur at the same time.

Some examples of **mutually exclusive** events are:

an individual being a son and a daughter simultaneously
a number that is an even whole number and a number between 0 and 1

> **Special Addition Rule**: If event A and event B are two **mutually exclusive** events, then the formula that describes this situation is:
>
> $P(\text{A or B}) = P(\text{A}) + P(\text{B})$

The probability of mutually exclusive events A or B happening is the sum of their respective probabilities. There is no need for the final term as in the **General Addition Rule** because the events **cannot** happen at the same time.

Note: Some books use $P(\text{A} \cup \text{B})$ instead of $P(\text{A or B})$ and $P(\text{A} \cap \text{B})$ for $P(\text{A and B})$.

EXAMPLE 5.5

(a) A box contains 10 discs numbered 1 through 10 inclusive. What is the probability of randomly selecting an even number **or** a number greater than seven?

(b) A fair die is rolled across a table. What is the probability of rolling an even number **or** a number less than 3?

(c) A weather reporting station determines that the probability of rain is 0.4, the probability of wind is 0.7, and the probability of rain and wind is 0.3. What is the probability that it will be rainy **or** windy?

(d) At a college, 45% of the students are women, 55% of the students are men. Also, 35% of the students are over 25 years of age, meaning 65% are 25 years of age or below. Forty percent of the women are over 25 years of age. What is the probability that a student is a woman **or** someone over 25?

(e) The table below shows the relationship among employees at an auto plant.

	MEN	WOMEN	TOTAL
MANAGEMENT	20%	20%	40%
LABOR	50%	10%	60%
TOTAL	70%	30%	100%

Find the probability of randomly selecting an employee that is in management **or** a woman.

(f) At a school meeting, 20% of those attending are fathers and 40% of those attending are mothers; the rest are children. What is the probability of randomly selecting a mother **or** a father from the group?

(g) A fair die is rolled. Find the probability of rolling a number that is odd **or** that is greater than 5.

SOLUTION 5.5

(a) The discs are numbered 1, 2, 3, 4, 5, 6, 7, 8, 9, 10. The probabilities are:

$$P(\text{A}) = P(\text{even}) = P(2, 4, 6, 8, 10) = \frac{5}{10} = \frac{1}{2}$$

$$P(\text{B}) = P(\text{number greater than } 7) = P(8, 9, 10) = \frac{3}{10}$$

$$P(\text{A and B}) = P(\text{even and greater than } 7) = P(8, 10) = \frac{2}{10}$$

Since the events are not mutually exclusive, we use

$P(\text{A or B}) = P(\text{A}) + P(\text{B}) - P(\text{A and B})$

$P(\text{even or} > 7) = P(\text{even}) + P(> 7) - P(\text{even and} > 7)$

$= \frac{5}{10} + \frac{3}{10} - \frac{2}{10} = \frac{8}{10} - \frac{2}{10} = \frac{6}{10} = \mathbf{\frac{3}{5}}$

(b) A die has six sides with numbers 1, 2, 3, 4, 5, 6. The probabilities are:

$P(\text{A}) = P(\text{even}) = P(2, 4, 6) = \frac{3}{6} = \frac{1}{2}$

$P(\text{B}) = P(\text{less than three}) = P(1, 2) = \frac{2}{6} = \frac{1}{3}$

$P(\text{A and B}) = P(\text{even and less than three}) = P(2) = \frac{1}{6}$

This is a **non-mutually exclusive event.**

$P(\text{A or B}) = P(\text{A}) + P(\text{B}) - P(\text{A and B})$

$P(\text{even or} < 3) = P(\text{even}) + P(< 3) - P(\text{even } \mathbf{and} < 3)$

$= \frac{3}{6} + \frac{2}{6} - \frac{1}{6} = \frac{5}{6} - \frac{1}{6} = \frac{4}{6} = \mathbf{\frac{2}{3}}$

(c) The probabilities in this problem are given as proportions. These proportions are a form that can be used to establish a probability (another form is that of a fraction as already shown). Here the given values are:

$P(\text{A}) = P(\text{rain}) = 0.4$

$P(\text{B}) = P(\text{windy}) = 0.7$

$P(\text{A and B}) = P(\text{rain and windy}) = 0.3$

Since the two events are **not mutually exclusive**, the needed formula is:

$P(\text{A or B}) = P(\text{A}) + P(\text{B}) - P(\text{A and B})$

$P(\text{A or B}) = P(\text{rain or windy})$
$= P(\text{rain}) + P(\text{windy}) - P(\text{rain and windy})$

$P(\text{A or B}) = 0.4 + 0.7 - 0.3 = 1.1 - 0.3 = 0.8$

(d) In this problem, the probabilities are given in percentages. We can easily convert these percentages to proportions by removing the % symbol and placing a decimal in front of the number.

In summary, we are finding the proportion of students who are women and who are over 25 years of age.

$P(\text{A}) = P(\text{women}) = 45\% = 0.45$

$P(\text{B}) = P(\text{over } 25) = 35\% = 0.35$

$P(\text{A and B}) = P(\text{woman and over } 25) = 40\% = 0.4$

A woman can be over 25 years of age, therefore the two sets are **not mutually exclusive**. The formula is:

$P(\text{A or B}) = P(\text{A}) + P(\text{B}) - P(\text{A and B})$

thus,

$P(\text{woman or over } 25) = P(\text{woman}) + P(\text{over } 25) - P(\text{woman and over } 25)$

$P(\text{A or B}) = 0.45 + 0.35 - 0.4 = 0.8 - 0.4 = 0.4$

(e) The probabilities are given in percentages. These percentages must be rewritten as proportions.

The probability of selecting an employee that is in management comes from the row total for management (40% or 0.4). The probability of selecting a woman is given in the column total under "women" (30% or 0.3). The probability of being a woman in management is 20% (0.2).

$P(\text{A}) = P(\text{woman}) = 0.3$

$P(\text{B}) = P(\text{management}) = 0.4$

$P(\text{A and B}) = P(\text{woman and manager}) = 0.2$

This problem has events which are **not mutually exclusive**. Thus,

$P(\text{A or B}) = P(\text{A}) + P(\text{B}) - P(\text{A and B})$

$P(\text{woman or management}) = P(\text{woman}) + P(\text{management}) - P(\text{woman and management})$

$P(\text{A or B}) = 0.3 + 0.4 - 0.2$

$= 0.7 - 0.2$

$= 0.5$

(f) Mothers and fathers are not overlapping sets. This problem involves events that are **mutually exclusive**.

The special rule applies.

$P(\text{A or B}) = P(\text{A}) + (\text{B})$

$P(\text{A}) = P(\text{mother}) = 0.4$

$P(\text{B}) = P(\text{father}) = 0.2$

$P(\text{A or B}) = P(\text{mother or father})$

$= P(\text{A}) + P(\text{B})$

$= 0.4 + 0.2$

$= 0.6$

(g) For a fair die, $P(\text{odd}) = P(A) = P(1, 3, 5) = \frac{3}{6} = \frac{1}{2}$

$P(> 5) = P(\text{B}) = P(6) = \frac{1}{6}$

There is **no** odd number greater than 5 and less than 6 on a die, therefore these are **mutually exclusive events**.

$P(\text{A or B}) = P(\text{A}) + P(\text{B})$

$P(\text{odd and greater than 5}) = P(\text{odd}) + P(> 5) = \frac{3}{6} + \frac{1}{6} = \frac{4}{6} = \frac{2}{3}$

The Multiplication Rule

When events happen in conjunction, the **multiplication rule** is used. There are two rules that apply to multiplication. The difference between the two rules is a matter of whether the events are **dependent** or **independent**.

Independent Events

> **Independent** events are events that do not affect the probability of each others occurring.

Some examples of **independent** events are:

a rainstorm in Montana and a festival in Greece

a dice roll and a card draw

> **Dependent** events are events that affect the probability of each other occurring.

Examples are:

a person buying a car and that person driving the car

a card drawn from a deck and a second card drawn from the same deck

> The multiplication rule for independent events A and B is:
>
> $$P(\text{A and B}) = P(\text{A}) \cdot P(\text{B})$$

The above rule says that the probability that event A **and** event B happen is the probability of event A happening **times** the probability of event B happening.

> The **multiplication rule** for **dependent events** A and B is:
>
> $$P(\text{A and B}) = P(\text{A}) \cdot P(\text{B} \mid \text{A})$$

The probability of both events A and B occurring is the probability of event A times the probability of event B happening **given** event A has already happened.

The formula could be written in the form:

$$P(\text{A and B}) = P(\text{B}) \cdot P(\text{A}|\text{B})$$

Note: P(A and B) could be written $P(\text{A} \cap \text{B})$ or P(AB).

EXAMPLE 5.6

(a) A container has 3 red cubes and 4 blue cubes. What is the probability that 2 red cubes are successively drawn? Assume that the cubes are not replaced after selection.

(b) A container has 3 red cubes and 4 blue cubes. What is the probability that 2 red cubes are drawn? Assume that once a cube is drawn, it is replaced.

(c) A game consists of spinning a dial with 5 colors and throwing a single die. The 5 colors are red, blue, yellow, green, and black. What is the probability of tossing an even number on the die and having the spinner stop on blue?

(d) A weather forecaster reports the probability of rain in Seattle is 0.6 and the probability of rain in Tampa is 0.5. What is the probability of rain in Seattle and Tampa? Assume independence.

(e) A supervisor in a large plan predicts that the probability of producing enough of item A is 0.3 and the probability of producing enough of item B is 0.6. What is the probability of producing enough of item A and **not** enough of item B? Assume independent events.

SOLUTION 5.6

(a) This is an example of a dependent situation. Once a cube is drawn, the sample space is reduced by one cube. Thus, the probabilities for drawing a second cube are changed as well.

There are 7 cubes in the original sample space. The probability of drawing a red cube, P(A), is $3/7$ (3 red out of 7 cubes). WE ASSUME WE'RE SUCCESSFUL IN DRAWING A RED CUBE.

Once the first red cube is drawn, the number of red cubes is now 2 out of a total of 6 cubes. Thus, $P(\text{B})$, the probability of drawing a second red cube, is $2/6$. The proper formula is:

$$P(\text{A and B}) = P(\text{A}) \cdot P(\text{B}|\text{A})$$

$$P(\text{A and B}) = P(\text{red and red}) = P(\text{red}) \cdot P(\text{red given red})$$

$$= \frac{3}{7} \cdot \frac{2}{6} = \frac{6}{7 \cdot 6} = \frac{1}{7}$$

(b) This problem is the same as the previous one (5.6a), except that the cube is **replaced** after it is drawn.

The replacement of the cube means that the total sample space and the number of outcomes of interest each remain the same for both selections.

The correct formula is the formula for **independent events**:

$$P(\text{A and B}) = P(\text{A}) \cdot P(\text{B})$$

$$P(\text{A}) = P(\text{red}) = \frac{3}{7}$$

$$P(\text{B}) = P(\text{red}) = \frac{3}{7}$$

$$P(\text{A and B}) = P(\text{red and red})$$

$$= P(\text{red}) \cdot P(\text{red})$$

$$= \frac{3}{7} \cdot \frac{3}{7}$$

$$= \frac{9}{49}$$

(c) Since the spinning of the dial does not affect the toss of the die, or vice versa, this is an independent situation.

$$P(\text{A and B}) = P(\text{A}) \cdot P(\text{B})$$

$$P(\text{A}) = P(\text{color blue}) = \frac{1 \text{ color}}{5 \text{ colors}} = \frac{1}{5}$$

$$P(\text{B}) = P(\text{even number on die}) = P(2, 4, 6) = \frac{3}{6} = \frac{1}{2}$$

$$P(\text{A and B}) = P(\text{blue and even number})$$

$$= P(\text{blue}) \cdot P(\text{even})$$

$$= \frac{1}{5} \cdot \frac{1}{2}$$

$$= \frac{1}{10}$$

(d) The problem states that these are independent events.

$$P(\text{A and B}) = P(\text{A}) \cdot P(\text{B})$$

$$P(\text{A}) = P(\text{rain in Seattle}) = 0.6$$

$$P(\text{B}) = P(\text{rain in Tampa}) = 0.5$$

$$P(\text{A and B}) = P(\text{rain in Seattle and rain in Tampa})$$

$$= P(\text{rain in Seattle}) \cdot P(\text{rain in Tampa})$$

$$= (0.6)(0.5) = 0.3$$

(e) This problem, again, is one of independence of events. However, we are looking for the complement of event B for our second probability. As mentioned earlier, the complement of event B has a probability that is one minus the probability of event B.

$$P(\text{B}') = 1 - P(\text{B})$$

In this case, the probability formula for independent events can be amended slightly to

$$P(\text{A and B}') = P(\text{A}) \cdot P(\text{B}')$$

$$P(\text{A}) = P(\text{event A}) = 0.3$$

$$P(\text{B}') = 1 - P(\text{event B}) = 1 - 0.6 = 0.4$$

$P(\text{A and B}') = P(\text{event A and not event B})$

$= P(\text{event A}) \cdot P(\text{not event B})$

$= (0.3)(0.4) = 0.12$

Some final notes about Addition and Multiplication Rules of probability.

For the **Addition Rule**:

The key word in several types of Addition Rule problems is "or."

Mutually exclusive events cannot happen at the same time (simultaneously).

For the **Multiplication Rule**:

The key words used in some problems of this type are "and," "then," or "and then."

If the words "with replacement" are used, it usually indicates a problem that has independent events.

If "without replacement" is used, it usually indicates dependent events.

5.7 CONDITIONAL PROBABILITY

In the previous section, we talked about using the multiplication formula:

$P(\text{A and B}) = P(\text{A}) \cdot P(\text{B} \mid \text{A})$

when selecting items **without** replacement. This is an example of the occurrence of one event affecting the probability of the occurrence of another event. The probability of one event is altered on the **condition** that another event has taken place.

The following illustration expands our look into conditional probability other than that demonstrated by "replacement" or "no replacement." Suppose we are using a deck of cards as our sample space. We are to determine the probability that we have selected a King of Hearts, when pulling one card from the deck of 52 cards. Normally, our probability would be $1/52$. However, if we learn that the card we drew was a "face" card, our probability rises to $1/12$ (since there are just 12 face cards in the deck).

As discussed earlier, we often assign events a letter designation for simplicity.

A = Drawing a "face" card.

B = Drawing a King of Hearts.

The probability of B is definitely changed when A occurs. Symbolically, we can note the two probabilities of B occurring whether A occurs or not.

$P(\text{B}) = \frac{1}{52}$ if we have no knowledge about A.

$P(\text{B}|\text{A}) = \frac{1}{12}$ if A has occurred.

$(\text{B}|\text{A})$ is read "B given A has occurred" or "B given A" or "B after A."

EXAMPLE 5.7

Each of the following examples involves two events, event A and event B. Determine the following probabilities:

$P(\text{A})$ $P(\text{B})$ $P(\text{B}|\text{A})$

(a) A = Drawing a "Club" card from a deck
B = Drawing a "3 of Clubs"from a deck

(b) A car dealer has a sales lot with an equal number of vans, sedan cars, economy cars, pick-up trucks, and luxury cars.

A = A customer selects a car.
B = A customer selects an economy car.

(c) A = Drawing a "Diamond" card from a deck.
B = Tossing a 3 on a die.

SOLUTION 5.7

(a) $P(A) = \frac{13}{52} = \frac{1}{4}$

There are 52 cards in a deck, with 13 of those cards labeled "Clubs."

$P(B) = \frac{1}{52}$

There is only one "3 of Clubs" in an entire deck of cards.

$P(B|A) = \frac{1}{13}$

Once we know that the card drawn is a "Club," then the sample space has been reduced to 13 cards. One card out of the 13 is a "3 of Clubs."

(b) $P(A) = \frac{3}{5}$

There are 3 types of cars out of 5 types of vehicles. Since the number of vehicles is the same in each category, we can select by category.

$P(B) = \frac{1}{5}$

There is one category of economy cars in the 5 types of vehicles.

$P(B|A) = \frac{1}{3}$

Given that there are only 3 types of cars and an economy car is one of the 3, the probability of B changes.

(c) $P(A) = \frac{13}{52} = \frac{1}{4}$

There are 13 diamonds in a deck of 52 cards.

$P(B) = \frac{1}{6}$

There is 1 three on a die out of 6 sides.

$P(B|A) = 1/6 = P(B)$

Rolling a die and selecting a card from a deck are independent. B is **not** dependent on A.

Using a Formula

The conditional probability of an event can be determined mathematically with the following formula:

> The conditional probability formula for events A and B is
>
> $$P(B|A) = \frac{P(\text{A and B})}{P(A)}$$

EXAMPLE 5.8

(a) The probability that it rains in St. Louis and Chicago is 0.25 and the probability that it rains in Chicago is 0.4. What is the probability that it rains in St. Louis given it rains in Chicago?

(b) Two dice are rolled and the sum of the dots of the upward faces is found. What is the probability that the sum is a 3, 6, or 8 given that the sum is odd?

SOLUTION 5.8

(a) A = Rain in Chicago
B = Rain in St. Louis

Define events A and B.

$P(A) = 0.4$
$P(\text{A and B}) = 0.25$

Use given information to assign probabilities.

$$P(B|A) = \frac{P(\text{A and B})}{P(A)}$$

Use the conditional probability formula to determine $P(B|A)$.

$$P(B|A) = \frac{0.25}{0.4} = 0.625$$

(b) Let A: There is an odd sum.
B: There is a sum of 3, 6, or 8.

Assign probabilties to the events A and B.

$$P(\mathrm{A}) = \frac{18}{36} = \frac{1}{2}$$

Out of the 36 possibilities, 18 sums are odd.

$$P(\mathrm{A \text{ and } B}) = P(3)$$

$$= \frac{2}{36} = \frac{1}{18}$$

When adding and finding the 36 possible sums, we find that there are only two ways to get a 3. (1 on die one and 2 on die two **or** 1 on die two and 2 on die one.)

$$P(\mathrm{B}|\mathrm{A}) = \frac{P(\mathrm{A \text{ and } B})}{P(\mathrm{A})}$$

Use the conditional probability formula to find $P(\mathrm{B}|\mathrm{A})$.

$$= \frac{\frac{1}{18}}{\frac{1}{2}} = \frac{1}{18} \cdot \frac{2}{1} = \frac{2}{18}$$

$$= \frac{1}{9}$$

5.8 BAYE'S THEOREM

Conditional probability looked at what happened to the probability of a later event, B, given that the event A had **already** occurred. We now look at the situation in the reverse. Can we determine the probability of an earlier event A, given that a later event B occurs? The answer is "yes" – if we can determine certain other probabilities.

This can be done with Baye's Formula.

Baye's Formula

Events $E_1, E_2 \ldots E_k \ldots E_n$ are all mutually exclusive events that make up the entire sample space. For any event A (with a non-zero probability):

$$P(E_k|A) = \frac{P(E_k) \cdot P(A|E_k)}{P(E_1) \cdot P(A|E_1) + \ldots + P(E_k) \cdot P(A|E_k) + \ldots + P(E_n) \cdot P(A|E_n)}$$

EXAMPLE 5.9

(a) Given two mutually exclusive events E_1 and E_2, with $P(E_1) = 0.7$, $P(E_2) = 0.3$. The probability, $P(A|E_1)$ is known to be 0.3 and $P(A|E_2)$ is 0.2. Find $P(E_1|A)$.

(b) The probability of rain in Tampa is 0.4, the probability of rain in London is 0.6. Given that it rains in Tampa, the probability of rain in Miami is 0.8. Given that it is raining in London, the probability of rain in Miami is 0.4. What is the probability of rain in Tampa, given that it is raining in Miami?

SOLUTION 5.9

(a) $P(E_1) = 0.7 \quad P(E_2) = 0.3$

$P(E_1|A) = 0.3$

$P(E_2|A) = 0.2$

List the assigned probabilities.

Write Baye's Formula in the format of this problem.

$$P(A|E_1) = \frac{P(E_1) \cdot P(A|E_1)}{P(E_1)P(A|E_1) + P(E_2)P(A|E_2)}$$

$$= \frac{(0.7)\ (0.3)}{(0.7)\ (0.3) + (0.3)\ (0.2)}$$

Substitute values and solve.

$$= \frac{0.21}{0.21 + 0.06}$$

$$= \frac{0.21}{0.27} = 0.78$$

(b) E_1 = Rain in Tampa
E_2 = Rain in London
A = Rain in Miami

Define each event according to E_1, E_2, and A.

$P(E_1) = 0.4$
$P(E_2) = 0.6$
$P(A|E_1) = 0.8$

$P(A|E_2) = 0.4$

Assign probability values.

Write Baye's Formula in the format of this problem.

$$P(E_1|A) = \frac{P(E_1) \cdot P(A|E_1)}{P(E_1)P(A|E_1) + P(E_2)P(A|E_2)}$$

$$= \frac{(0.4)\ (0.8)}{(0.4)\ (0.8) + (0.6)\ (0.4)}$$

Substitute the values and solve for the answer.

$$= \frac{0.32}{0.32 + 0.24}$$

$$= \frac{0.32}{0.66} = 0.48$$

Practice Exercises

1. (a) A camera shop sells 3 types of camera bodies, 4 lenses and 2 flash attachments. How many ways can the camera store display a combination of flash, lense, and camera body?

   (b) A construction company has 30 electricians, 15 plumbers, and 43 laborers. If the company wishes to send a 3-person team consisting of an electrician, a plumber, and a laborer, how many possible ways can the team be formed?

2. (a) A coach has 20 people trying for 12 positions on a track team. How many ways can the coach select the players, if order is not important?

   (b) Eight people are running in an election to qualify for 5 positions on the fire district board. How many combinations are possible? (Order is not important.)

3. (a) Eight hockey teams run a race for 1st, 2nd, and 3rd places. How many ways can the teams finish 1st, 2nd, or 3rd?

   (b) A comedy club has 12 comedians on this evenings show. One will be selected as a winner in the amature competition. How many ways can the selection turn out?

4. Find the probability of the following simple events.

   (a) There are 5 parrots and 3 parakeets in a group of 8 birds. Find the probability of randomly selecting a parakeet.

   (b) A box contains 3 red cubes and 12 yellow cubes. What is the probability of randomly selecting a green cube?

   (c) A container has 6 red cubes, 3 yellow cubes, and 8 blue cubes. What is the probability of drawing a yellow cube?

5. (a) A box contains 10 discs numbered 1 through 10 inclusive. What is the probability of randomly selecting an odd number **or** a number less than 6?

   (b) A fair die (one dice cube) is rolled across a table. What is the probability of rolling an even number **or** a number greater than 5?

   (c) A weather reporting station determines that the probability of snow is 0.8, the probability of wind is 0.3, and the probability of snow and wind is 0.1. What is the probability that it will be snowy **or** windy?

6. (a) A container has 8 red cubes and 4 blue cubes. What is the probability that 2 red cubes are drawn? Assume that the cubes are not replaced after selection.

   (b) A container has 3 red cubes and 4 blue cubes. What is the probability that 3 red cubes are drawn on 3 tries? Assume that once a cube is drawn, it is replaced.

7. Each of the following examples have 2 events, event A and event B. Determine the following probabilities:
   $P(A)$ $P(B)$ $P(B|A)$

   (a) A = Drawing a "Heart" card from a deck
   B = Drawing a "King of Hearts" from a deck

   (b) A furniture company has a condominium section that contains a sofa, a recliner chair, and an end table. There are 4 other types of sections in the furniture store as well.

   A = A customer selects the condominium package.
   B = A customer selects the reclining chair.

Answers

1. (a) 24

   (b) 19,350

2. (a) 125,970

   (b) 56

3. (a) 336

   (b) 12

4. (a) 3/8

   (b) 0

   (c) 3/17

5. (a) 7/10

   (b) 1/2

   (c) 1

6. (a) 7/18

   (b) 27/343

7. (a) $P(A) = 1/4$, $P(B) = 1/52$, $P(B|A) = 1/3$

   (b) $P(A) = 1/5$, $P(B) = 1/7$, $P(B|A) = 1/3$

6

Statistics

Statistics is a body of knowledge that concerns the theory and methods for describing, explaining, drawing inferences, and making conclusions under conditions of uncertainty.

A **sample** is any subset of a population.

Types of Statistics

When we analyze data from samples, we can use techniques from two types of statistics, **inferential statistics** and **descriptive statistics.**

> **Inferential statistics** involves the generalization of information about a sample to a larger population.
>
> **Descriptive statistics** is the process of describing the characteristics of a sample without further generalization.

We will be discussing descriptive statistics in this chapter. This includes displaying data, measures of central tendency, and measures of dispersion.

6.1 DISPLAYING QUALITATIVE DATA

Qualitative Data

> Data such as race, sex or occupation that occurs in discrete categories is termed **qualitative** data.

This type of data is usually displayed in bar graphs (sometimes called column graphs or histograms) or circle graphs (also known as pie charts).

Many samples have items that can be placed in categories. Examples of this are the color of candies in a bag, the types of goods manufactured by a country, a state-by-state list of the amount of coal produced, or the number of World Series wins by team.

> The number of items in each category is called the **frequency** of that category.

The charts or graphs used to represent the data must display the relative frequencies of the data. The height of the bar or the size of the slice demonstrates the relationship between various amounts in each category of data being graphed.

Figure 6.1 shows a pie chart (circle graph) that displays the categories of cars by type that have been found in a particular parking lot. The numbers in each slice are the frequencies.

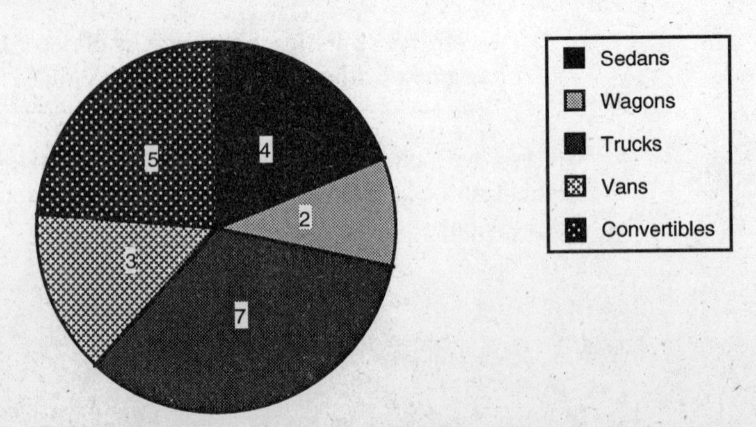

Figure 6.1 Pie Chart of Cars in a Parking Lot

Matching the number inside the slice with the box shading to the left of the car type, we can build the following table.

car type	frequency
sedan	4
wagons	2
trucks	7
vans	3
convertibles	5

Table 6.1 Types of Cars

Figure 6.2 shows this same data in a **bar graph**. The frequency is on the vertical axis. Note that the top of the bar represents the number of cars in that particular category.

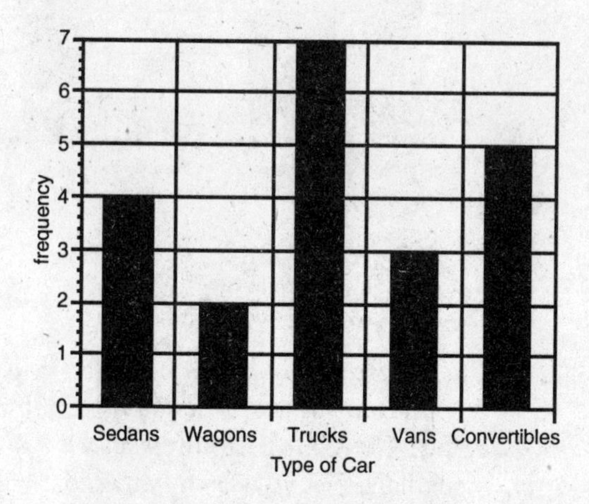

Figure 6.2 Bar Graph for Type of Car

Again, the bar graph is for **qualitative** or **categorical data.**

EXAMPLE 6.1

Draw a bar graph to demonstrate the following sample:
The types of fish in a small pond are noted. There are 8 bass, 3 catfish,

2 bluegill, and 5 sunfish.

SOLUTION 6.1

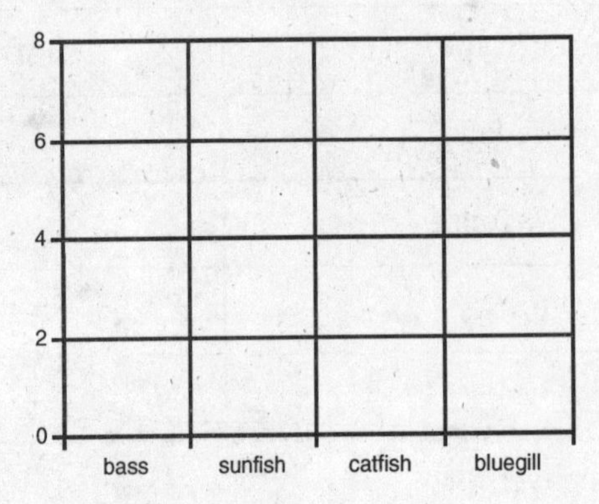

Draw the axes, with the vertical axis as the frequency. The highest frequency is the maximum value on the vertical axis. The types of fish goes on the horizontal axis.

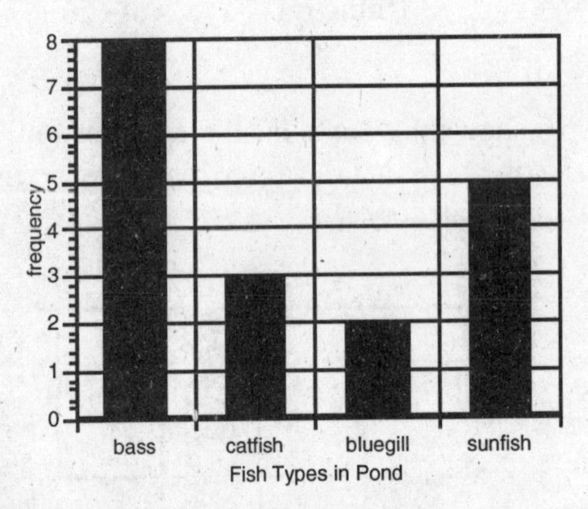

Make the height of each bar rise to the appropriate frequency shown on the vertical axis. Label the graph.

EXAMPLE 6.2

Draw a pie graph representing the following data:

Children in an elementary school are selécted, and their hair color is noted. There are 8 children selected in all. Three have blonde hair, 2 have black hair, 2 have brown hair, and 1 has red hair.

SOLUTION 6.2

Since there are 8 elements in the sample, we can divide the circle into 8 parts. This will work for any multiple of 8 as well.

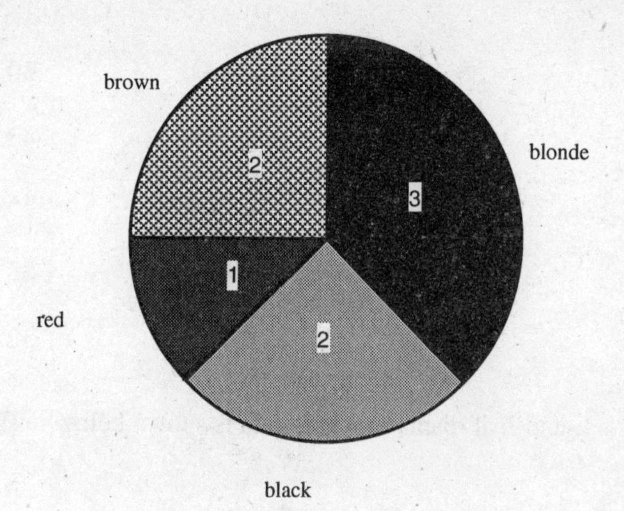

There are 3 children with blonde hair, so we shade 3/8 of the circle. We shade 2/8 for brown hair. The last 2 categories, red and black, complete the graph.

6.2 DISPLAYING QUANTITATIVE DATA

Quantitative Data

Quantitative data is derived from samples that have continuous numerical data. This data can assume any value within a given range.

Examples of quantitative data are height, weight, IQ, or income. The display of quantitative data is usually done with a frequency distribution. The frequency distributions discussed in this section are the **histogram** and the **stem-and-leaf** display. The **normal curve** is another frequency distribution that will be discussed later in the chapter.

Consider the following data representing miles per gallon in gasoline consumption for 20 cars.

14.2	17.3	27.5	30.2
25.6	10.2	40.1	45.7
31.5	35.2	29	36
43	42.6	31	23.6
27.2	31	34.6	52.3

The stem-and-leaf display for this data is shown below in Figure 6.3.

stem	leaf						
1	0.2	4.2	7.3				
2	7.5	5.6	9	3.6	7.2		
3	0.2	1.5	5.2	6.0	1.0	4.6	1.0
4	0.1	5.7	3.0	2.6			
5	2.3						

Figure 6.3 Stem-and-Leaf Display

The stem (in this case) is the "tens" digit. The "ones" digit along with the "tenths" digit form the leaf. The display affords a "quick" look at how the data is distributed for the sample.

You might note that the leaves are not in numerical order. They need not be.

The histogram for this data would look like Figure 6.4.

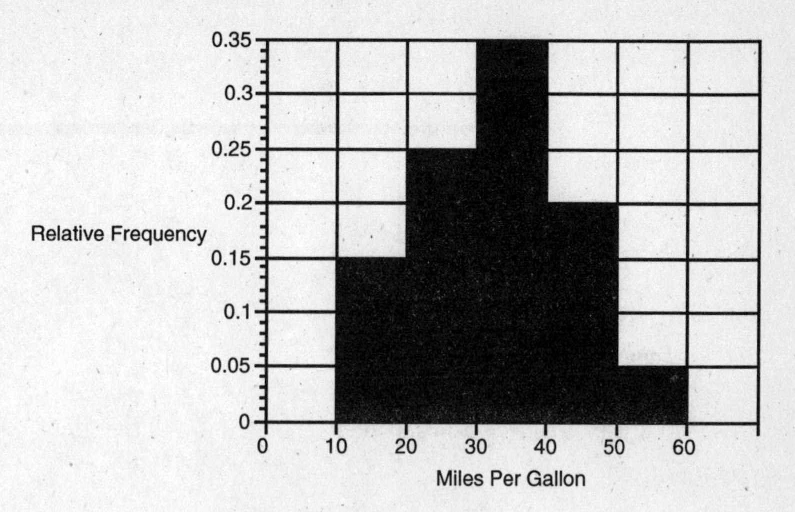

Figure 6.4 Histogram

The histogram in Figure 6.4 has 5 classes (bars). Each class ***must*** have the ***same width*** and the numbers shown on the horizontal axis are called the **class boundaries**. No data values (**scores**) can fall on the class boundaries.

The vertical axis for this particular histogram has **relative frequencies**. A relative frequency is the number of members of a certain class divided by the total number in the entire sample. A relative frequency can not be greater than one or less than zero. You might recall that there are similar boundaries for a probability. Relative frequencies are sometimes used to determine probabilities.

EXAMPLE 6.3

For the following sample data construct a stem-and-leaf display.

1.5	2.6	3.1	1.7
3.3	2.1	1.8	2.0
5.1	4.3	3.9	3.7

SOLUTION 6.3

stem	leaf			
1	0.5	0.8	0.7	
2	0.6	0.1	0	
3	0.1	0.3	0.9	0.7
4	0.3			
5	0.1			

We first draw the vertical and horizontal lines typical of the stem-and-leaf display. In this example, the data has "ones" and "tenths." We choose to make the "ones" the stem, and the "tenths" the leaf.

EXAMPLE 6.4

For the following scores, construct a stem-and-leaf display.

117	123	145	137
127	119	113	129
133	136	142	111
107	161	149	118

SOLUTION 6.4

stem	leaf				
10	7				
11	7	9	3	1	8
12	3	7	9		
13	7	3	6		
14	5	2	9		
15					
16	1				

Here the T-shaped table of the stem-and-leaf display is set up. In this case, the stem is the "hundreds" and "tens" digits. The leaves are the "ones" digits. Note that although we have a 15 stem, it has no leaf. There was no data value in the 150's.

EXAMPLE 6.5

Set up a relative frequency histogram for the following sample scores.

10	25	36	17
30	27	39	26
41	33	19	22
58	49	43	30

SOLUTION 6.5

A histogram usually contains from 5 to 12 classes (these numbers have been established by custom and efficiency of application). The number of classes chosen within the range is up to the individual drawing the histogram. For this example, we will use 5 classes, since the sample is small. The procedure for the histogram follows:

highest score – lowest score	
58 – 10 = 48	Subtract the smallest score from the largest score to find the **range**.
$\frac{\text{range}}{\text{number of classes}} = \frac{48}{5} = 9.6$ => 10 (rounding up)	Divide the range by the number of classes to obtain a **class width**. Round up to the next larger whole number.
Start at 9 + 10 = 19 + 10 = 29 + 10 = 39 5 classes + 10 = 49 + 10 = 59	Start the first class just below the smallest number, 10 (in this case use 9). Add the class width to this number and to each successive class boundary until you have 5 classes.

class	class width
1	9.5-19.5
2	19.5-29.5
3	29.5-39.5
4	39.5-49.5
5	49.5-59.5

We cannot have a data value fall on a class boundary. Note that the data points 19 and 39 fall on class boundaries (the 19 and 39 of the previous step). We can easily fix this. Just go to any digit one power of 10 less than the smallest power of ten you have in the data. In this case, if we make 9.5, 19.5, 29.5, 39.5, 49.5, and 59.5 the class boundaries, we can have no possibility of a piece of data falling on the boundary. No data value has a 10th place.

class	class width	freq.
1	9.5-19.5	3
2	19.5-29.5	4
3	29.5-39.5	5
4	39.5-49.5	3
5	49.5-59.5	1

List the number of data points in each class.

class	class width	freq.	rel.freq.
1	9.5-19.5	3	3/16 = .18
2	19.5-29.5	4	4/16 = .25
3	29.5-39.5	5	5/16 = .32
4	39.5-49.5	3	3/16 = .18
5	49.5-59.5	1	1/16 = .06
			total = .99

Find the relative frequency by dividing each class frequency by the total number of data points (in this case 16).

The total of the relative frequencies should approximately equal 1.

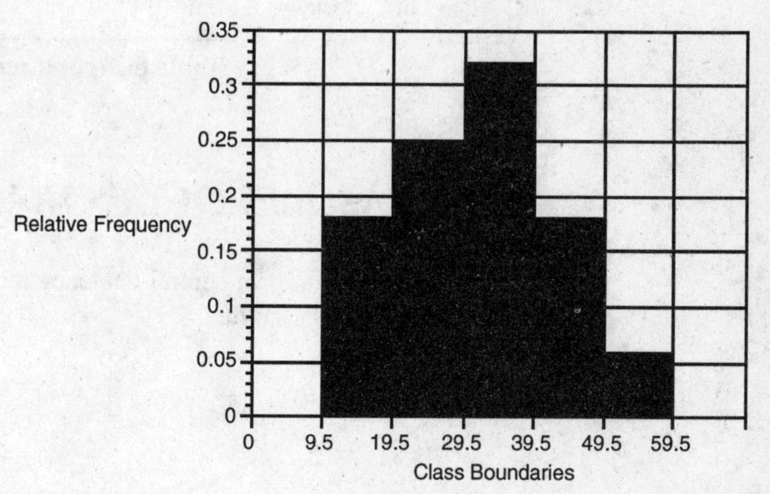

Plot the histogram by listing classes on the horizontal axis, and the relative frequency on the vertical axis. Since the highest relative frequency is .32, there is no need to go much higher.

It is important to know the similarities and differences between a bar (column) graph and a histogram. Table 6.2 shows these relationships.

item	**bar graph**	**histogram**
type of data	qualitative (categorical)	quantitative (continuous)
data as frequencies	used more often here	used less often here

Table 6.2 Difference Between Bar Graph and Histogram

item	bar graph	histogram
data as relative frequencies	used less often here	used more often here
bars touching	almost never	always
class boundaries	no	yes

Table 6.2 (continued)

6.3 MEASURES OF CENTRAL TENDENCY

The three measures of central tendency most often used are the **mean**, the **median**, and the **mode**.

The Mode

> The **mode** (if there is one) is the most frequently occurring value. There may be more than one mode to a sample.

EXAMPLE 6.6

Find the mode of the following samples.

(a) 3, 3, 7, 2, 3, 4, 8, 9

(b) 4, 1, 8, 2, 6, 5, 10, 11

(c) 3, 4, 4, 4, 1, 2, 7, 7, 7

SOLUTION 6.6

(a) The data value that occurs most often is 3. It is the mode.

(b) No data value occurs most often; there is no mode.

(c) There are two scores that occur most often. These scores are 4 and 7. This a **bimodal** sample.

The Median

> The **median** is the middle data value within ordered data, after the scores have been arranged in numerical order.

EXAMPLE 6.7

Find the median of the following samples.

(a) 3, 4, 7, 5, 3, 6, 2, 1, 1

(b) 4, 5, 2, 6, 1, 7

SOLUTION 6.7

(a) If there are an odd number of scores, we arrange the sample in numerical order. We then pick the score that is literally in the middle of all the scores. In this sample, there are 9 scores, thus we pick the middle value.

1, 1, 2, 3, 3, 4, 5, 6, 7

The median is 3.

(b) In this sample, there are an even number of data values. We arrange the values into numerical order. The order can be from highest to lowest, or from lowest to highest.

7, 6, 5, 4, 2, 1

↑↑

middle two values

When the sample has an even number of values, we average the middle two values to get the median. The average is $\frac{4+5}{2} = 4.5$.

Therefore, 4.5 is the median.

The Mean

> The **mean** (sometimes called the **arithmetic mean**) is the sum of all the data values divided by the number of data points in the sample. The letter, ***n***, is often used to denote the number of data points.

EXAMPLE 6.8

Find the mean of the following samples.

(a) 4, 6, 2, 3, 1, 7, 6, 8

(b) 2.1, 3.2, 1.1, 2.7, 5.3

SOLUTION 6.8

(a) We divide the sum of the data values by the number of data values, n. The value of n for this sample is 8.

$$\frac{4+6+2+3+1+7+6+8}{8} = \frac{37}{8} = 4.625$$

We use the symbol $\bar{x}$ for the mean. $\bar{x}$ is 4.625.

(b) The mean (average) for $n = 5$ scores is:

$$\bar{x} = \frac{2.1+3.2+1.1+2.7+5.3}{5} = \frac{14.4}{5} = 2.88$$

Note: The mean is the measure of central tendency **most** affected by extreme scores in the sample. The relationships between the mean, median, and mode relating to frequency distributions is discussed in section, "The Normal Curve."

6.4 MEASURES OF DISPERSION

The measures discussed in the previous section show where the data tends to fall near some intermediate value. There is also another way to look at a sample. Both the sample 1, 2, 3, 4, 5 and the sample 0, 0, 3, 0, 12 have n = 5 and the mean = 3, but it is apparent that they are very different. This is why measures of dispersion are important items as well.

The Range

Perhaps the easiest measure of dispersion to evaluate is the **range**. This was mentioned briefly in the section, "Displaying Qualitative Data," as it was used to find class boundaries.

> The **range** is the difference between the largest data value and the smallest data value in a sample.

EXAMPLE 6.9

Find the range of the following samples:

(a) 4, 7, 11, 23, 3, 7, 18, 20

(b) 2.1, 4.5, 3.2, 7.1, 5.3, 2.7, 3.4, 5.1

SOLUTION 6.9

(a) 4, 7, 11, **23**, **3**, 7, 18, 20 — List sample. Note extreme values.

23 – 3 = 20 = **range** — The range is the highest value minus the smallest value.

(b) **2.1**, 4.5, 3.2, **7.1**, 5.3, 2.7, 3.4, 5.1 — List sample. Note extreme values.

$7.1 - 2.1 = 5 =$ **range** — The range is the highest value minus the smallest value.

Variance and Standard Deviation

Two other measures of dispersion are related very closely to each other, and it is easy find one after the other is known.

> The **variance,** s^2, is the square of the average deviation about the mean.
>
> The difference between a data value and the mean is known as the **deviation score.**
>
> The **standard deviation** is the square root of the variance. It is the measure of the average deviation about the mean.

The examples presented shortly will show two ways in which we can get the standard deviation and the variance.

The formula for a sample **variance** is:

$$s^2 = \frac{\sum (x - \bar{x})^2}{n - 1}$$

$\sum$ means "sum"

The formula for the **standard deviation** is:

$$s = \sqrt{\frac{\sum (x - \bar{x})^2}{n - 1}}$$

EXAMPLE 6.10

Find the variance and standard deviation of the following samples.

(a) 2, 3, 1, 6, 3

(b) 7, 3, 4, 4, 5

SOLUTION 6.10

(a) The variance is found through a stepwise process. The first thing found is the mean. The sum of 2 + 3 + 1 + 6 + 3 is 15. This is divided by the n-value 5 to get the mean of 3.

score	dev. score	(dev score)^2
2		
3		
1		
6		
3		

The **scores** (elements) are listed in the left-most column. Find their sum.

$\sum x = 15$

score	dev. score	(dev score)^2
2	$2 - 3 = -1$	
3	$3 - 3 = 0$	
1	$1 - 3 = -2$	
6	$6 - 3 = 3$	
3	$3 - 3 = 0$	

The deviation score is each score minus the mean. The **sum of the deviation scores,** $\sum (x - \bar{x})$, **should always be zero.**

$\sum x = 15 \quad \sum (x - \bar{x}) = 0$

score	dev. score	(dev score)^2
2	$2 - 3 = -1$	1
3	$3 - 3 = 0$	0
1	$1 - 3 = -2$	4
6	$6 - 3 = 3$	9
3	$3 - 3 = 0$	0

The deviation score for each is squared. The sum of the deviation scores squared is the numerator of the variance.

$\sum x = 15 \quad \sum (x - \bar{x}) = 0 \quad \sum (x - \bar{x})^2 = 14$

The number of data points is 5. Therefore, $n - 1 = 5 - 1 = 4$.

The variance is $\frac{14}{4}$ $= \mathbf{3.5} = s^2$. The **square root** of the **variance** is the **standard deviation**. In this case, $\sqrt{3.5} \approx 1.87 = s$.

(b) There is an alternate formula that is sometimes used as a "shortcut." It is as follows.

$$s^2 = \frac{\sum x^2 - \frac{(\sum x)^2}{n}}{n-1}$$

We need only square the scores, and square the sum of the scores.

x	x^2
7	49
3	9
4	16
4	16
5	25

We make a column for the data values. They are summed ($\sum x = 23$ here). The x-squared values are summed here in the second column.

$\sum x = 23$ $\sum x^2 = 115$ $n = 5$

$$s^2 = \frac{115 - \frac{23^2}{5}}{5-1} = \frac{115 - 529/5}{4} = \frac{115 - 105.8}{4} = \frac{9.2}{4} = 2.3$$

When the variance is 2.3, the standard deviation is $\sqrt{2.3} \approx 1.52$.

Note: The standard deviation of the above sample is 1.52. An approximation for the standard deviation is the range divided by 4.

The range in the above sample is 7 – 3 = 4. Therefore, 4/4 is equal to one, which is not too far from 1.52. The method of dividing the range by 4 should be used as an approximation only.

6.5 THE NORMAL CURVE

Normal Distributions

The **normal curve** is a frequency distribution that occurs in many natural systems. Many populations and samples are called **normal** when they follow the pattern of this curve. Figure 6.5 shows the normal curve.

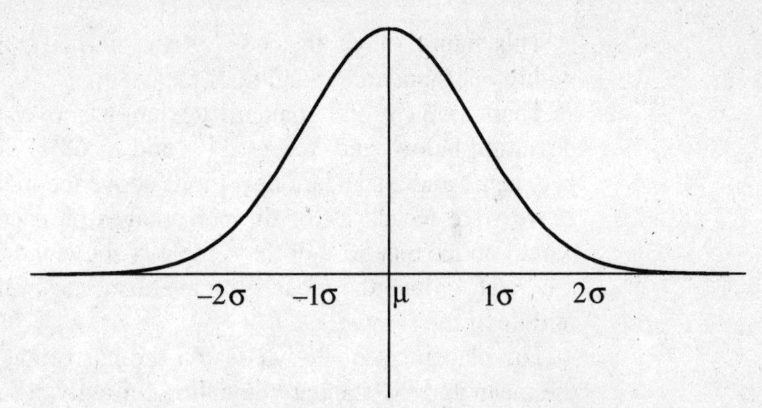

Figure 6.5 The Normal Curve

The center line through the curve shows where mean value is located. The values to the right and the left of the center line are one and two standard deviations above and below the mean. The σ is the **population symbol** for the standard deviation. The *s* is used for the **sample symbol** for standard deviation.

The normal curve has percentages derived for each standard deviation above the mean and below the mean. Figure 6.6 has the percentages between each standard deviation on the curve.

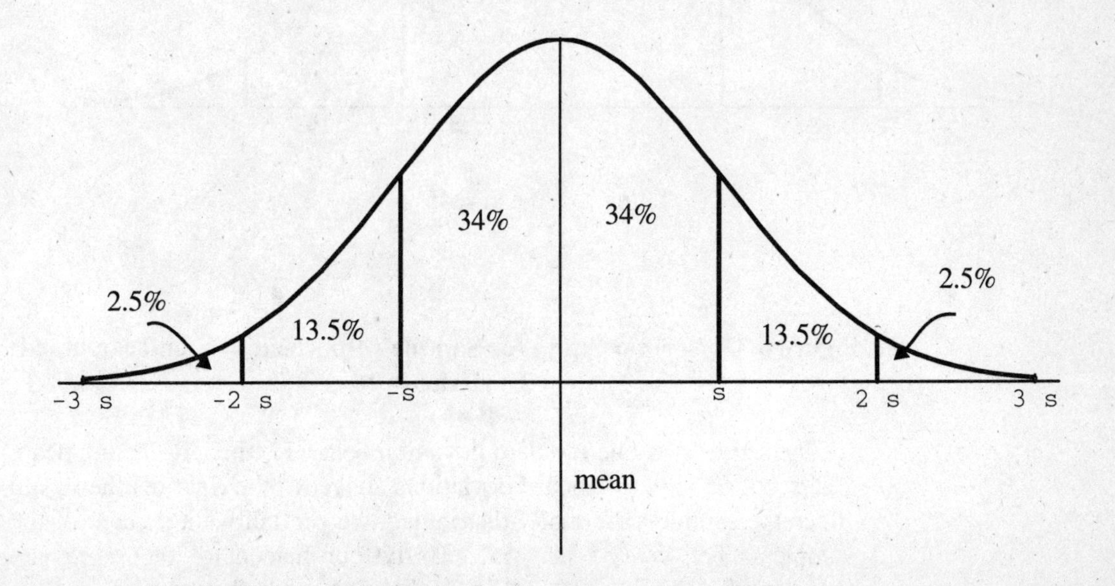

Figure 6.6 The Normal Curve with Percentages

This figure shows that 68% of the normal population (or sample) is within one standard deviation of the mean.

There is 34% one standard deviation above, and 34% one standard deviation below, and 34% + 34% add to 68%. There is another 13.5% between 1 and 2 standard deviations above the mean and below the mean. If we were to add all of the percentages on each side of the mean, we would notice that 50% of the sample is above the mean and 50% is below the mean. Since the mean is an average, we would expect it to be in the middle of the sample.

The placement of the values on the horizontal axis is found by using the mean and the standard deviation for a given sample. Suppose a sample had a mean of 45 and a standard deviation 10.

The curve would look like Figure 6.7.

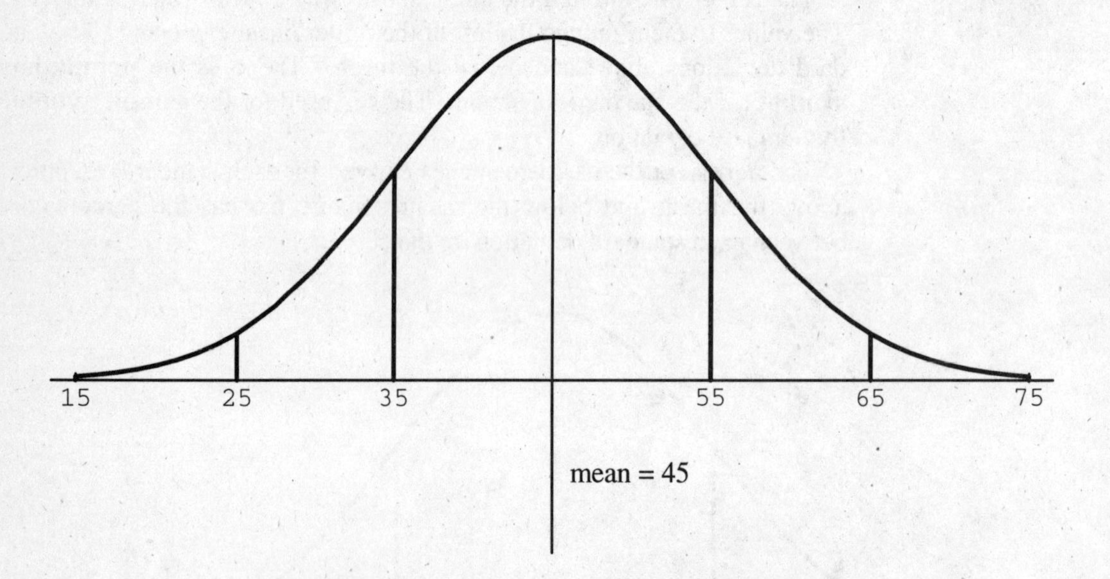

Figure 6.7 Normal Curve for Sample with Mean = 45 and Standard Deviation = 10

The score 35 is one standard deviation below (to the left of) the mean. The score 65 is two standard deviations above (to the right of) the mean. Since the sample is normally distributed, we can tell what percent of the sample is between 35 and 65. Thirty-four percent of the scores lie between 35 and 45. 34% + 13.5% of the scores lie between 45 and 65. When these percentages are added, we get 34% + 34% + 13.5% = 81.5%. Therefore, 81.5% of the scores fall between 35 and 65.

EXAMPLE 6.11

Using the normal distribution, find the percentages requested for the following samples.

(a) A sample has a mean of 100 and a standard deviation of 15; find the percentage of scores that lie between 70 and 115.

(b) A sample has a mean of 10 and a standard deviation of 2; find the percentage of data values between 12 and 14.

(c) A sample has a mean of 40 and a variance of 16; find the percentage of scores between 32 and the mean.

SOLUTION 6.11

(a) The sample has a mean of 100. We use this as a base from which to subtract the standard deviation (15) and to which to add the standard deviation. The subtraction and addition is as follows:

mean – 3 standard deviations = 100 – 3(15) = 100 – 45 = 55

mean – 2 standard deviations = 100 – 2(15) = 100 – 30 = 70

mean – 1 standard deviations = 100 – 1(15) = 100 – 15 = 85

mean + 1 standard deviations = 100 + 1(15) = 100 + 15 = 115

mean + 2 standard deviations = 100 + 2(15) = 100 + 30 = 130

mean + 3 standard deviations = 100 + 3(15) = 100 + 45 = 145

When the curve's horizontal axis is complete, the results are as follows:

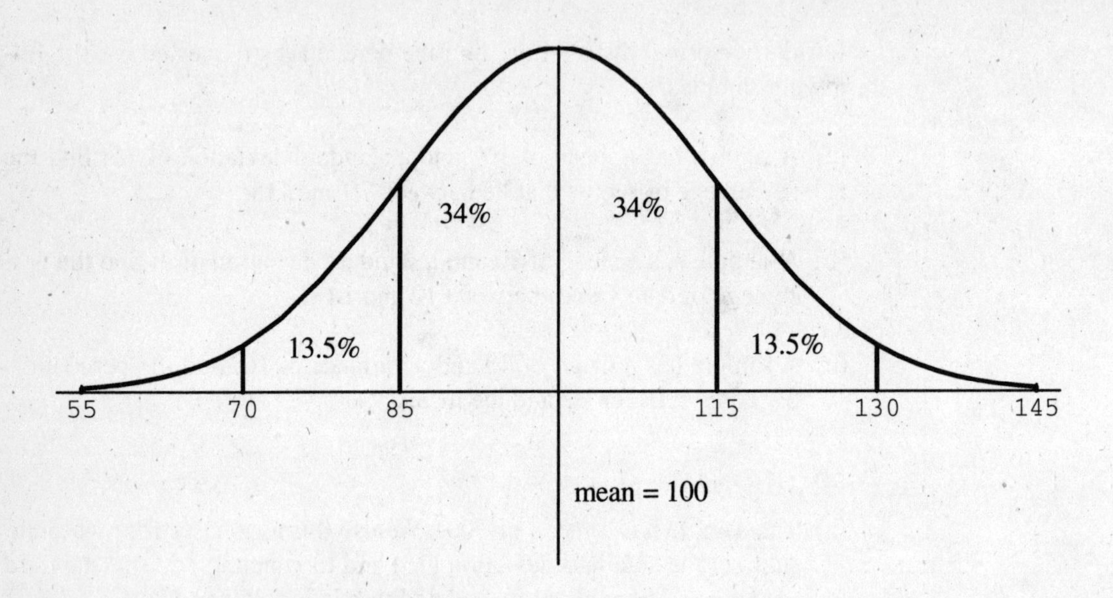

We now look for the scores of 70 and 115 and add the percentages between those scores. The shaded area in the figure below shows the subset of the sample that is of interest.

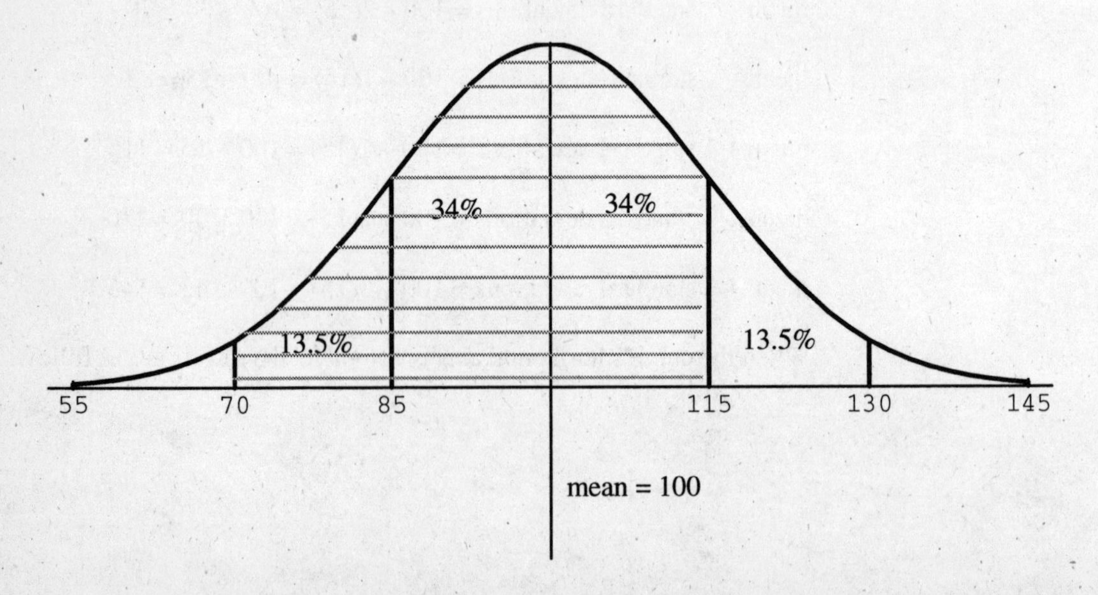

By adding the percentages in the shaded area: 13.5% + 34% + 34%, we see that 81.5% of the scores are between 70 and 115.

b) The sample has a mean of 10. Use this as a base from which to subtract the standard deviation (2) and to add to the standard deviation. The subtraction and addition is as follows:

mean – 3 standard deviations = 10 – 3(2) = 10 – 6 = 4

mean – 2 standard deviations = 10 – 2(2) = 10 – 4 = 6

mean – 1 standard deviations = 10 – 1(2) = 10 – 2 = 8

mean + 1 standard deviations = 10 + 1(2) = 10 + 2 = 12

mean + 2 standard deviations = 10 + 2(2) = 10 + 4 = 14

mean + 3 standard deviations = 10 + 3(2) = 10 + 6 = 16

When the curve's horizontal axis is complete, the results are as follows:

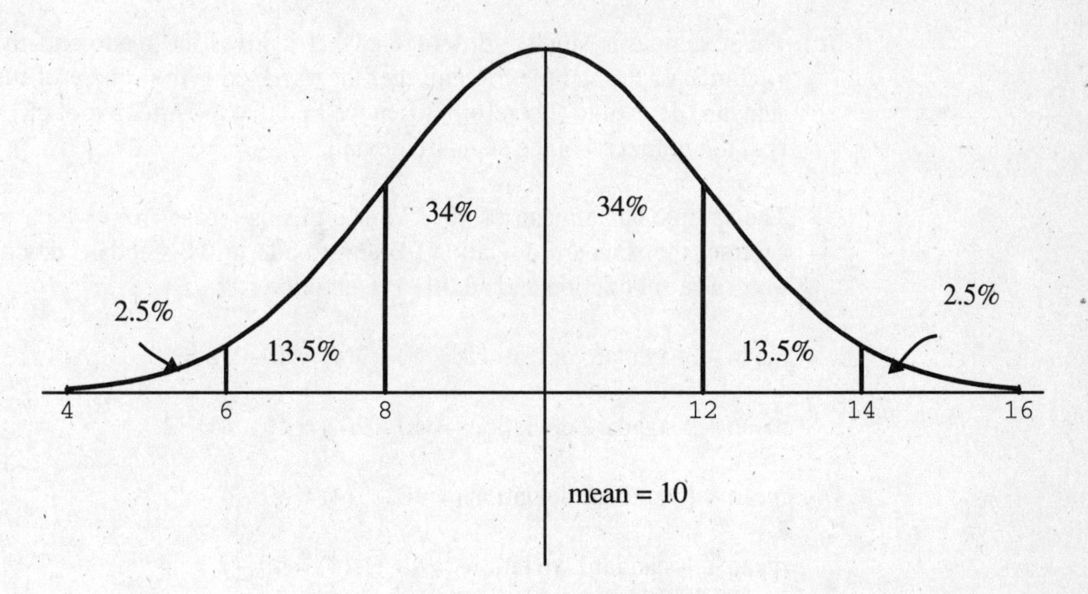

The data values between which we are to calculate the percentages are 12 and 14. The normal curve modified for this sample shows that to be 13.5%. The shaded area below indicates the area under consideration.

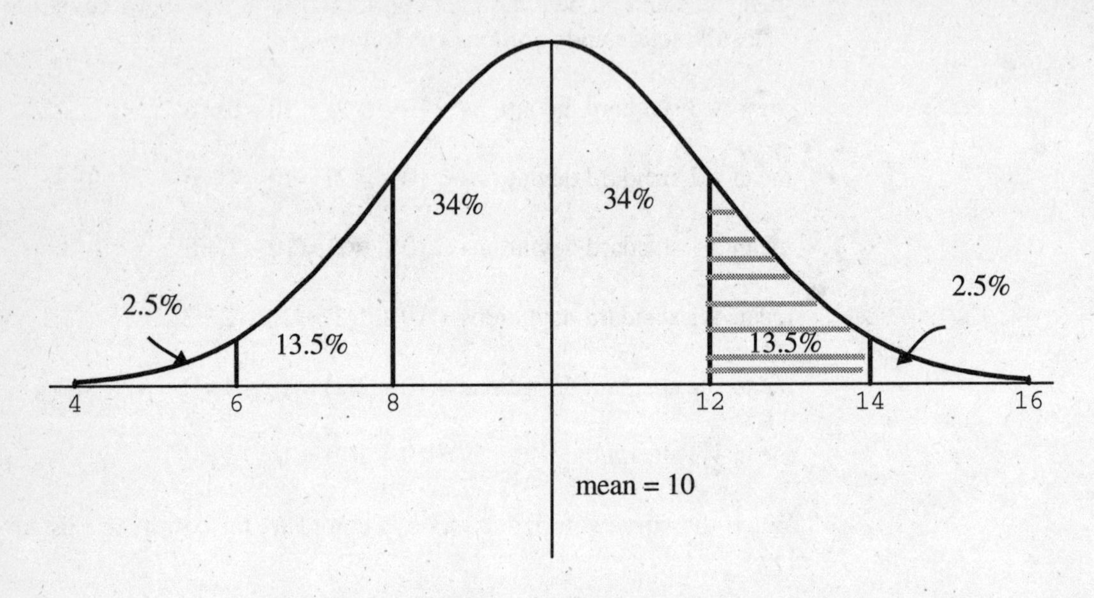

(c) This example is slightly different in that it gives the mean and the **variance** of the sample. Recall that the variance is the square of the standard deviation. Therefore, when we find that the square root of 16 is 4, the number 4 is the standard deviation.

The sample has a mean of 40. We use this as a base from which to subtract the standard deviation (4) and to add to the standard deviation. The subtraction and addition is as follows:

mean – 3 standard deviations = 40 – 3(4) = 40 – 12 = 28

mean – 2 standard deviations = 40 – 2(4) = 40 – 8 = 32

mean – 1 standard deviations = 40 – 1(4) = 40 – 4 = 36

mean + 1 standard deviations = 40 + 1(4) = 40 + 4 = 44

mean + 2 standard deviations = 40 + 2(4) = 40 + 8 = 48

mean + 3 standard deviations = 40 + 3(4) = 40 + 12 = 52

When the curve's horizontal axis is complete, the results are as follows:

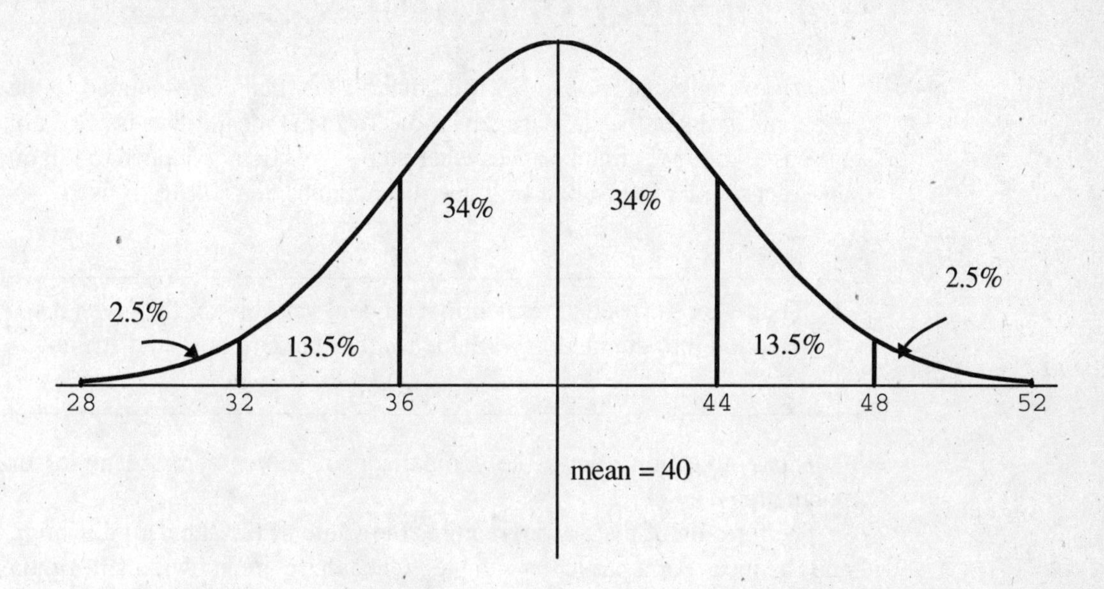

The data values between which we are to calculate the percentages are 32 and 40. The normal curve modified for this sample shows the percentage to be 13.5% + 34 % = 47.5%. The shaded area below indicates the area under consideration.

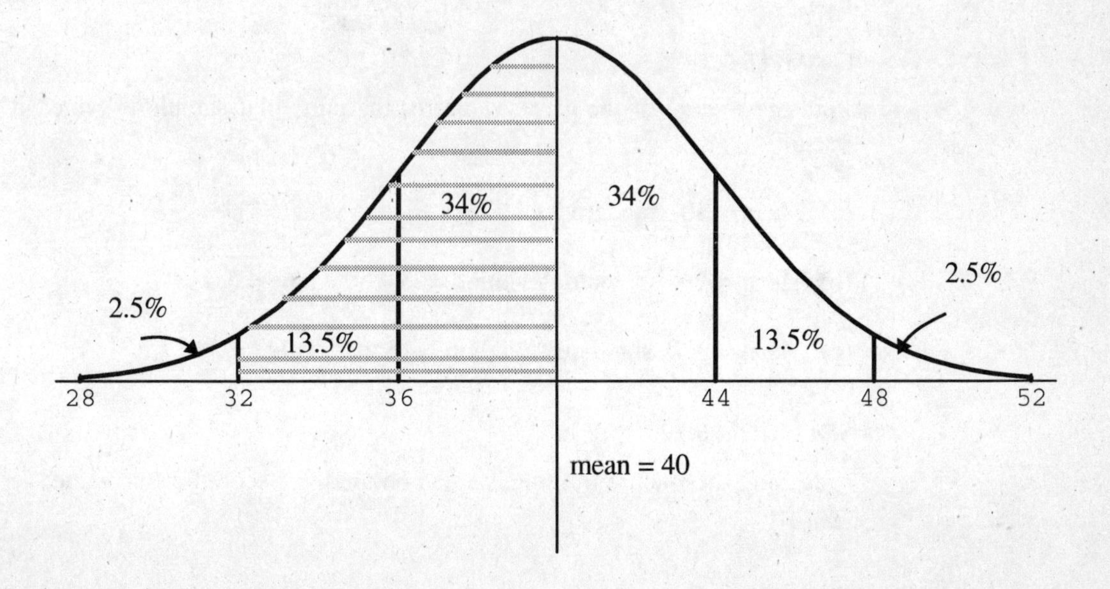

6.6 MEASURE OF POSITION: Z-SCORE

The examples in section 6.5 used data values that corresponded to the exact positions of the standard deviations from the mean. This is often not the case. In those circumstances where the scores do not happen to fall on an exact standard deviation from the mean values, the **z-score** is used.

> The **z-score** is used to transform a random variable for the given distribution into the random variable for the standard normal distribution.

This conversion allows the comparison of scores across samples or populations.

The formula of the z-score requires the value of the standard deviation and the mean for a particular sample. Once these are known, the formula below can be used to standardize a value from a sample.

$$z = \frac{x - \bar{x}}{s}$$

Here, x is the score, **x-bar** is the mean, and s is the standard deviation.

EXAMPLE 6.12

Find the z-score of the given value for the particular sample in which it occurs.

(a) Mean = 30, standard deviation = 8, data value = 34.

(b) Mean = 1.8, standard deviation = .3, data value = 1.4.

(c) Mean = 270, standard deviation = 18, data value = 304.

SOLUTION 6.12

(a) Using the z-score fromula, we can obtain the z-score for this value:

$$z = \frac{x - \bar{x}}{s}$$

$$= \frac{34 - 30}{8} = \frac{4}{8} = \frac{1}{2} = 0.5$$

The meaning of this ***z*-score** is that the value 34 from this sample is **.5** standard deviation **above** the mean.

(b) Using the z-score formula, we can obtain the z-score for this value:

$$z = \frac{x - \bar{x}}{s}$$

$$= \frac{1.4 - 1.8}{0.3} = \frac{-0.4}{0.3} = -\frac{0.4}{0.3} = -1.33$$

The meaning of this ***z*-score** is that the value 1.4 from this sample is **1.33** standard deviations **below** the mean.

(c) Using the z-score formula, we can obtain the z-score for this value:

$$z = \frac{x - \bar{x}}{s}$$

$$= \frac{304 - 270}{18} = \frac{34}{18} = \frac{17}{9} = 1.88$$

The meaning of this ***z*-score** is that the value 304 from this sample is **1.88** standard deviations **above** the mean.

6.7 DISCRETE RANDOM VARIABLES

There are many situations that occur which can have only certain numerical values as a result. When a single die is rolled, the dot pattern facing up can have one through six dots. A grade school can have as many as eight grades, with each grade labelled by a whole number. Elevators stop at whole-numbered floors. House addresses are not given in irrational number values (such as, $\sqrt{3}$).

If we were to randomly select a dot pattern, a grade, a floor number, or an address, we would expect to obtain a limited number of values. Since these values do not form a smooth continuum, we call these values **discrete.**

> A **discrete random variable** is a unique numerical value selected, at random, from a limited set of values in certain popluations.

Each of the discrete random variables is assigned a probability value, which– as we discussed earlier– can be values from 0 to 1. The probability assigned each discrete random variable is an indication of the relative abundance of the variable within the population.

The value of the discrete variable and the probability of occurrence within a population are often tabulated as in the following example. Each random variable comes under the heading of "x", with each probability designated by $P(x)$.

i	1	2	3	4
x_i	1	3	5	6
$P(x_i)$	.1	.5	.2	.2

where i is label for each x_i.

Discrete Random Variable

Note that all the decimals on the bottom row add to 1.0 (.1 + .5 + .2 + .2 = 1.0). This must be true for the entire population. The values for x can be any real number, but the probability values, $P(x)$, can only be positive values less than or equal to 1. A probability of zero, while allowed mathematically, is not needed.

EXAMPLE 6.13

Determine whether each of the following tables represent a true population of discrete random variables. If not, tell why not.

(a)

i	1	2	3
x_i	0	2	4
$P(x_i)$	.3	.2	.1

(b)

i	1	2	3	4
x_i	−3	−1	2	5
$P(x_i)$	.1	.6	.2	.1

(c)

i	1	2	3
x_i	$\frac{\pi}{2}$	π	$\frac{3\pi}{2}$
$P(x_i)$	.3	.4	.6

(d)

i	1	2	3
x_i	4	10	72
$P(x_i)$	−.1	.5	.6

SOLUTION 6.13

(a) This is *not* a table of the entire population. The probabilities total .6 (.3 + .2 + .1). Note that the values for x (0, 2, 4) are allowed.

(b) This is a true population of discrete random variables. (.1 + .6 + .2 + .1 = 1.0).

(c) This is *not* a true population of discrete random variables. The probabilities add to more than 1 (.3 + .4 + .6 = 1.1).

(d) This is *not* a true population of discrete random variables. There is a *negative* number (−.1) represented as a probability.

Expected Value

Once a true population has been established, we can find the equivalent of the mean of this population, the **expected value**.

> The **expected value**, E(X), is the average of the probability values for all the members of the population.

The mathematical formula for the expected value is :

$$E(X) = \sum x_i \cdot P(x_i)$$

for each x_i in the population. Here, x_i is a particular discrete random variable. $P(x_i)$ is the probability of occurence for that variable. $E(X)$ is the expected value. Σ is the mathematical symbol for summation.

The process of finding the expected value is demonstrated by the following.

i	1	2	3
x_i	2	5	6
$P(x_i)$	.3	.4	.3

Since the probabilities are all positive and sum to 1, we can treat this as a valid population and find the expected value.

The expected value is found through the following calculations.

$E(X) = \sum x_i \cdot P(x_i)$ for all i.

$$\begin{aligned} E(X) &= x_1 \cdot P(x_1) + x_2 \cdot P(x_2) + x_3 \cdot P(x_3) \\ &= 2\ (.3) \quad + 5\ (.4) \quad + 6\ (.3) \\ &= .6 \quad + \quad 2.0 \quad + \quad 1.8 \\ &= 4.4 \end{aligned}$$

The mean (or expected value) is 4.4. This is the average value of the entire sample taken from such a population. Note that the value of 4.4 is *not* a discrete data value in this particular population. In fact, we do not often find the expected value to be one of the discrete random variables in the population. We just "expect" the average of a given set of randomly chosen numbers sampled from this population to be around 4.4.

EXAMPLE 6.14

Find the expected value, $E(X)$, for each of the following populations.

(a)

i	1	2	3	4
x_i	5	8	11	12
$P(x_i)$	.1	.4	.4	.1

(b)

i	1	2	3
x_i	–1	0	2
$P(x_i)$	.6	.3	.1

(c)

i	1	2	3
x_i	.5	1.5	2.5
$P(x_i)$	.3	.3	.4

SOLUTION 6.14

(a)

The expected value, $E(X)$, is given by $\sum x_i \cdot P(x_i)$.

i	1	2	3	4
x_i	5	8	11	12
$P(x_i)$	.1	.4	.4	.1

$$E(X) = \sum x_i \cdot P(x_i) = x_1 \cdot P(x_1) + x_2 \cdot P(x_2) + x_3 \cdot P(x_3) + x_4 \cdot P(x_4)$$

Expand the sum.

$E(X) = 5(.1) + 8(.4) + 11(.4) + 12(.1)$	Substitute the numerical values.
$E(X) = 0.5 + 3.2 + 4.4 + 1.2$	Multiply each pair.
$E(X) = 9.3$	Add to find the expected value.
(b)	The expected value, $E(X)$, is $\sum x_i \cdot P(x_i)$.

i	1	2	3
x_i	–1	0	2
$P(x_i)$	.6	.3	.1

$E(X) = \sum x_i \cdot P(x_i) = x_1 \cdot P(x_1) + x_2 \cdot P(x_2) + x_3 \cdot P(x_3)$	Expand the sum.
$E(X) = (-1)(.6) + (0)(.3) + (2)(.1)$	Substitute the numerical values.
$E(X) = -0.6 + 0 + 0.2$	Multiply the values.
$E(X) = -0.4$	Sum to find the expected value.
(c)	The expected value, $E(X)$, is $\sum x_i \cdot P(x_i)$.

i	1	2	3
x_i	.5	1.5	2.5
$P(x_i)$	.3	.3	.4

$E(X) = \sum x_i \cdot P(x_i) = x_1 \cdot P(x_1) + x_2 \cdot P(x_2) + x_3 \cdot P(x_3)$	Expand the sum.

$E(X) = (.5)(.3) + (1.5)(.3) + (2.5)(.4)$	Substitute the numerical values.
$E(X) = (.15) + (.45) + (1)$	Multiply the values.
$E(X) = 1.6$	Sum to find the expected value.

Variance and Standard Deviation of a Probability Distribution

The population of a discrete distribution also has a variance and, hence, a standard deviation. When we talked about variance, we stated that it was the sum of the deviation scores, $(x - \bar{x})$, squared and divided by the sample size.

In the case of a discrete population known as a **discrete probability distribution**, the sample mean, $\bar{x}$, must be replaced by a population mean, μ. In a probability distribution, the population mean is known as the expected value, $E(X)$. This was determined using the techniques of the previous section.

Applying the symbol change, the formula for the variance looks like

$$V(X) = \frac{\sum (x_i - E(X))^2}{N}$$

$V(X)$ is the symbol for variance of a probability distribution.
$E(X)$ is the expected value.
N is the size of the population.
x is the value of a discrete random variable.
$\sum$ is the symbol for summation.

We can use the probabilty, $P(x_i)$, associated with each discrete random variable, x_i, to group the proportion for each x.

The formula now becomes,

$$V(X) = \sum (P(x_i))(x_i - E(X))^2$$

where $P(x_i)$ is the probability is associated with the discrete random variable, x_i. Algebraically, this can be rewritten,

$$V(X) = \left(\sum (x_i)^2 P(x_i)\right) - (E(X))^2$$

The following table shows a probability distribution. This table will help us demonstrate how to determine the variance and standard deviation of the probability distribution.

i	1	2	3
x_i	0	2	4
$P(x_i)$	.3	.4	.3

The expected value, $E(x)$, is determined by summing the product of each discrete random variable and its probability.

$$E(X) = \sum x_i P(x_i) = x_1 P(x_1) + x_2 P(x_2) + x_3 P(x_3)$$

$$= 0(.3) + 2(.4) + 4(.3) = .8 + 1.2 = 2$$

The variance, $V(X)$, is determined by next finding the sum of each discrete random variable squared times its probability. Then the square of the expected value is subtracted.

$$V(X) = \left(\sum x^2 \cdot P(x_i)\right) - (E(X))^2$$

$$= x_1^2 P(x_1) + x_2^2 P(x_2) + x_3^2 P(x_3) - (E(x))^2$$

$$= 0^2 (0.3) + 2^2 (0.4) + 4^2 (0.3) - (2)^2$$

$$= 0 + 1.6 + 4.8 - 4$$

$$= 2.4$$

The standard deviation is the square root of the variance.

$$\sigma = \sqrt{V(X)} = \sqrt{2.4} \approx 1.55$$

EXAMPLE 6.15

Find the variance and standard deviation of the following probability distributions.

(a)

i	1	2
x_i	5	10
$P(x_i)$	.8	.2

(b)

i	1	2	3
x_i	2	3	6
$P(x_i)$	.1	.8	.1

SOLUTION 6.15

x	$P(x)$	$x \cdot P(x)$	x^2	$x^2 P(x)$

A more efficient method of determining the variance and standard deviation of a probability distribution uses a table with the headings at the left.

(a)

x	$P(x)$	$x \cdot P(x)$	x^2	$x^2 P(x)$
5	.8			
10	.2			

Fill in the first two columns of the table from the given problem.

x	$P(x)$	$x \cdot P(x)$	x^2	$x^2 P(x)$
5	.8	.4	25	20
10	.2	2	100	20

Determine x^2 and find the products of $x \cdot P(x)$ and $x^2 \cdot P(x)$.

x	$P(x)$	$x \cdot P(x)$	x^2	$x^2P(x)$
5	.8	.4	25	20
10	.2	2	100	20
		$\sum = 2.4$		$\sum = 40$

Add the columns headed $x \cdot P(x)$ and $x^2P(x)$.

$$E(X) = \sum x \cdot P(x) = 2.4$$

$E(X)$ is 2.4.

$$V(X) = \left(\sum x^2P(x)\right) - (E(x))^2$$

The square of $E(X)$ is subtracted from the sum of the last column.

$$V(X) = 40 - (2.4)^2$$
$$V(X) = 40 - 5.76 = 34.24$$

$V(X) = 34.24$

$$\sigma = \sqrt{V(X)} = \sqrt{34.24} \approx 5.85$$

The standard deviation is the square root of the variance.

$V(X) = 34.24$ $\quad\sigma \approx 5.85$

(b)

x	$P(x)$	$x \cdot P(x)$	x^2	$x^2P(x)$
2	.1			
3	.8			
6	.1			

Fill in the first two columns of the table for the given problem.

x	$P(x)$	$x \cdot P(x)$	x^2	$x^2P(x)$
2	.1	.2	4	.4
3	.8	2.4	9	.72
6	.1	.6	36	21.6

Determine x^2 and find the products of $x \cdot P(x)$ and $x^2 \cdot P(x)$.

x	$P(x)$	$x \cdot P(x)$	x^2	$x^2P(x)$
2	.1	.2	4	.4
3	.8	2.4	9	7.2
6	.1	.6	36	3.6
		$\sum = 3.2$		$\sum = 11.2$

Find the sum of columns under $x \cdot P(x)$ and $x^2P(x)$.

$$E(X) = \sum x \cdot P(x) = 3.2$$

$E(X)$ is 3.2.

$$V(X) = \left(\sum x^2 P(x)\right) - (E(X))^2$$

The variance is equal to the sum of the last column minus $E(X)$ squared.

$$V(X) = 11.2 - (3.2)^2$$
$$V(X) = 11.2 - 10.24 = 0.96$$

$V(X)$ is 0.96.

$$\sigma = \sqrt{V(X)} = \sqrt{0.96} \approx 0.98$$

The standard deviation is the square root of the variance.

$V(X) = 0.96$ $\sigma = 0.98$

6.8 BINOMIAL DISTRIBUTION

There are many examples of populations that have only **two** types of outcomes. These outcomes are usually referred to as "success" and "failure." This arises in situations such as those described below.

(a) A question with only "yes" or "no" answers.

(b) Respondents select one item from two choices.

(c) Votes "for" or "against" an initiative.

(d) An object works or "fails" to work.

(e) A switch is "on" or "off."

These are termed binomial (or "two-name") populations. No matter what the original context of the population is, we let p equal the probability of "success" and q equal the probability of "failure."

We term a "success" the probability of interest and a "failure" the probability that the situation of interest does not happen. Since there are only two variables in the population (p and q), the probabilities p and q must sum to 1.

$$p + q = 1$$

$$p = 1 - q$$

The variables p and q are the probabilities of success and failure. We can form a table where X(success) = 1 and X(failure) = 0 with p equal to the probability of success and q equal to the probability of failure. This is done since "yes," "no," "success," "failure," etc. do not have numerical values. If the probability of success is .6 and failure = $1 - p$ = .4, the table looks as below:

	p	q
value	1	0
probability	.6	.4

Suppose we were looking at all samples of 3 p's and 2 q's from this population. The mean of this type of sample is found using a process similar to that of finding the expected value $E(X)$. The mean for this type of sample is:

$$\mu = 1(0.6) + 1(0.6) + 1(0.6) + 0(0.4) + 0(0.4)$$
$$= .6 + .6 + .6 = 1.8$$

If the sample does not include the entire population, $\bar{x}$ is used instead of μ. The terms using the q value are really not necessary for the final answer since $q = 0$. Thus, for binomially distributed populations

$$\mu = np$$

where n is the sample size and p is the probability of success.

The variance for a sample from a binomial population is

$$s^2 = npq$$

where n is the sample size.
p is the probability of "success."
q is the probability of "failure."

The standard deviation of a binomial population sample can be shown to be

$$s = \sqrt{npq}$$

(The symbols n, p, q are as described as above.)

EXAMPLE 6.16

Find the mean, variance, and standard deviations of the following samples.

(a) The sample size is 20, the probability of "success" is .3, the probability of "failure" is .7.
(b) The sample size is 100, the probability of "success" is .75.

SOLUTION 6.16

(a) $n = 20$ $p = .3$ $q = .7$	List the known values.
$\bar{x} = np$	The mean is found using $\bar{x} = np$.
$\bar{x} = 20\,(0.3) = 6$	Substitute and multiply to find the mean.
$s^2 = npq$	The variance equation is $s^2 = npq$.
$= (20)(.3)(.7)$ $= 6(.7) = 4.2$	The product of the substituted values is the variance.
$s = \sqrt{s^2} = \sqrt{npq} = \sqrt{4.2} \approx 2.05$	The standard deviation is the square root of the variance.
(b) $n = 100$ $p = .75$ $q = 1 - .75 = .25$	List the known values. q is $1 - p$.
$\bar{x} = np$	The mean is found using $\bar{x} = np$.
$\bar{x} = (100)\,(0.75)$ $= 75$	Substitute and multiply to find the mean.
$s^2 = npq$	The variance equation is $s^2 = npq$.
$= (100)(.75)(.25)$	The product of the substi-

$= 18.75$ — tuted values is the variance.

$s = \sqrt{s^2} = \sqrt{npq} \approx 4.33$ — The standard deviation is the square root of the variance.

EXAMPLE 6.17

(a) A company produces electrical circuit breakers for home use. From past experience, the company has found that 1% of the circuit breakers produced do not work correctly after manufacture. What is the average number of failures the company can expect with a production run of 10,500 circuit breakers?

(b) A county in Wyoming has a voter make-up of 60% Republican and 40% Democrat. Assume a sample of 10 people were randomly selected and 7 of those people were Republican. How many standard deviations above the mean is this number of Republicans in a 10 person sample?

SOLUTION 6.17

(a) $n = 10{,}500$ — List n, p, and q.
$p = 0.01$ (from 1%)
$q = 1 - p = 1 - 0.01 = 0.99$

Note: At this time, we must emphasize that, even though p is defined as the probability of "success," the percent failure of the circuit breakers is assigned to p. The "success" referred to in the standard definition can also be applied to the focus of the problem statement. Hence, we might think of p as the "success of finding a failed circuit breaker."

$\mu = np$ — The mean formula is used to find the solution.

$\mu = (10{,}500)(0.01)$

$\mu = 105$

Thus, we would expect 105 failed circuit breakers "on the average" to occur in 10,500 units produced.

(b) $n = 10$ — List the known values.
$p = 0.6$ (from 60%)
$q = 0.4$ (from 40%)

$x = 7$ (where x is the number of "successes" in this *particular* sample)

Note: The probability of success refers to the focus of the problem statement, and does not relate to your political preference.

$\bar{x} = np = 10(0.6) = 6$ The mean is found.

$$s = \sqrt{npq} = \sqrt{10\,(0.6)\,(0.4)}$$

$$= \sqrt{10\,(0.024)}$$

$$= \sqrt{0.24} \approx 0.49$$

The standard deviation is 0.49.

To find out how many standard deviations above the mean 7 is, we use the z-score formula.

$$z = \frac{x - \mu}{\sigma} \text{ where } \sigma = 0.49,\ \mu = 6,\ x = 7$$

$$= \frac{7 - 6}{0.49} = \frac{1}{0.49} = 2.04$$

Thus, the score of 7 is 2.04 standard deviations above the mean.

Practice Exercises

1. Draw a bar graph to demonstrate the following sample:
   The types of cars in a small used car lot are noted. There are six Chevrolets, three Fords, four Toyotas, and five Chryslers.

2. Draw a pie graph representing the following data:
   Expenditures for a household budget are $150 for food, $600 for housing, $200 for transportation, $50 for miscellaneous.

3. For the following sample data construct a stem-and-leaf display.

55	103	97	84
78	72	61	77
87	101	59	96

4. Set up a relative frequency histogram for the following samples scores.

10	25	36	17
30	27	39	26
41	33	19	22
58	49	43	30

5. Find the mode of the following samples.

   (a) 8, 8, 5, 4, 3, 8, 7

   (b) 3, 4, 8, 1, 0, 2, 9, 6, 7

   (c) 4, 4, 4, 5, 8, 8, 9, 7, 6, 10, 23

6. Find the median of the following samples.

   (a) 3, 4, 7, 5, 3, 6, 2, 1, 1

   (b) 4, 5, 2, 6, 1, 7

7. Find the mean of the following samples.

   (a) 10, 20, 34, 25, 36

   (b) .1, .25, .34, .45, .12

8. Find the range of the following samples:

   (a) 3, 5.6, 7, 3, 6.7, 23, 1.2

   (b) 102, 345, 296, 123, 564, 387

9. Find the variance and standard deviation of the following samples.

   (a) 1, 4, 4, 3, 2

   (b) 6, 2, 9, 1, 2

10. Using the normal distribution, find the percentages requested for the following samples.

    (a) A sample has a mean of 80 and a standard deviation of 10; find the percentage of scores that lie between 70 and 100.

    (b) A sample has a mean of 1.5 and a standard deviation of.25; find the percentage of data values between 1.75 and 2.00.

    (c) A sample has a mean of 20 and a variance of 9; find the percentage of scores between 26 and the mean.

11. Find the z-score of the given value for the particular sample in which it occurs.

    (a) Mean = 12, standard deviation = 1, data value = 9.

    (b) Mean = 2.7, standard deviation = .4, data value = 3.

    (c) Mean =1543, standard deviation = 170, data value = 1217.

12. Determine whether each of the following tables represents a true population of discrete random variables. If not, tell why not.

    (a)

i	1	2	3
x_i	0	2	5
$P(x_i)$	.7	.2	.1

    (b)

i	1	2	3	4
x_i	0	1	4	10
$P(x_i)$	.1	-.2	.2	.9

13. Find the expected value, $E(x)$, for each of the following populations.

    (a)

i	1	2	3	4
x_i	2	6	10	11
$P(x_i)$	.1	.3	.4	.2

    (b)

i	1	2	3
x_i	−2	0	2
$P(x_i)$	.2	.6	.2

14. Find the variance and standard deviation of the following probability distributions.

    (a)

i	1	2
x_i	2	4
$P(x_i)$	.6	.4

    (b)

i	1	2	3
x_i	-1	0	2
$P(x_i)$	.1	.3	.6

15. Find the mean, variance, and standard deviations of the following samples.

    (a) The sample size is 10, the probability of "success" is .4, the probability of "failure" is .6.

    (b) The sample size is 1000, the probability of "success" is .63.

Answers

1.

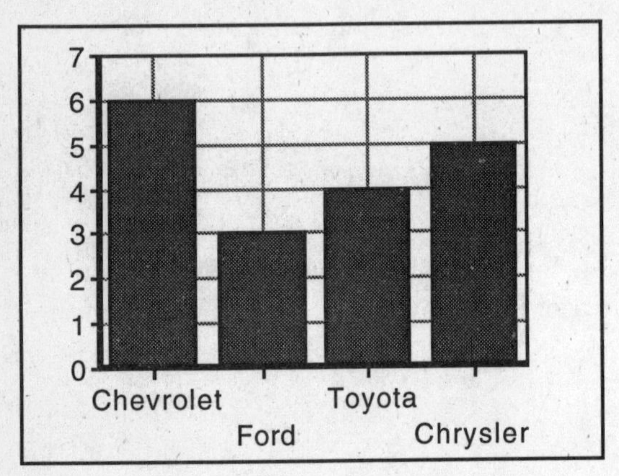

2.

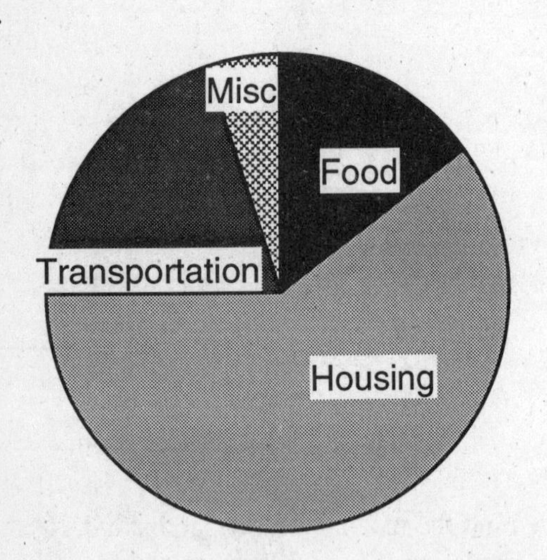

3.

Stem	Leaf
5	5, 9
6	1
7	2, 7, 8
8	4, 7
9	6, 7
10	1, 3

4.

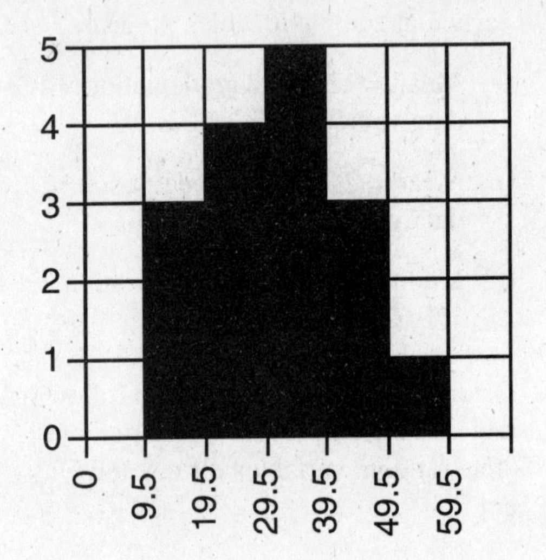

5. (a) 3

   (b) none

   (c) 4

6. (a) 3

   (b) 4.5

7. (a) 25

   (b) .252

8. (a) 20

   (b) 462

9. (a) 1.304, 1.7

   (b) 3.391, 11.5

10. (a) 81.5%

    (b) 13.5%

    (c) 47.5%

11. (a) -3

    (b) $-.75$

    (c) -1.917

12. (a) True, all probabilities sum to 1, all positive.

    (b) False, probabilities do not sum to 1, some negative.

13. (a) 8.2

    (b) 0

14. (a) $s = 1.033$, $s^2 = 1.0667$

    (b) $s = 1.1972$, $s^2 = 1.43333$

15. (a) $\mu = 4$; $s^2 = 2.4$; $s = 1.549$

    (b) $\mu = 630$, $s^2 = 233.1$, $s = 15.26$

7

Functions

This chapter introduces the notation, language and concepts used to discuss functions. Several broad categories of functions will be introduced in this chapter, and subsequently used in our study of calculus.

7.1 FUNCTIONS AND GRAPHING FUNCTIONS

Functions are used to describe relationships between variables. Furthermore, we insist that each input value produce exactly one output value.

A function is a rule that associates one and only one output value with each input value.

The input values are called the **domain** of the function, and the output values are called the **range** of the function.

Function Notation

We generally denote functions by lower or upper case italic letters such as f, g, F, or G. If we use x to represent the input value, we use $f(x)$, read f of x, to represent the output value. For example, $f(x) = 2x$ means for each input value of x, the output value is two times the input value.

"Find $f(3)$" means find the value of the output when the input is 3. For $f(x) = 2x$,

$$f(3) = 2(3) = 6$$

EXAMPLE 7.1

If $g(x) = 3x^2 + 2x - 1$ find

(a) $g(0)$

(b) $g(-2)$

(c) $g(x+h)$

SOLUTION 7.1

(a) $g(0) = 3(0)^2 + 2(0) - 1$ Substitute $x = 0$.

$= 0 + 0 - 1$ Simplify.

$= -1$

(b) $g(-2) = 3(-2)^2 + 2(-2) - 1$ Substitute $x = -2$.

$= 3(4) - 4 - 1$ Simplify exponents before multiplying.

$= 12 - 4 - 1$ Multiply.

$= 7$

(c) $g(x+h) = 3(x+h)^2 + 2(x+h) - 1$ Substitute $x = x + h$.

Recall:

$(x+h)^2 = (x+h)(x+h)$

$= x^2 + 2xh + h^2$

$= 3x^2 + 6xh + 3h^2 + 2x + 2h - 1$ Use the distributive property.

Piecewise Functions

Some relationships are more appropriately described using piecewise functions. For example, an employer may pay \$6 per hour for the first 40 hours an employee works. Overtime pay is \$9 per hour for each hour worked over 40 hours. If x represents the number of hours worked in a week, then,

$$f(x) = \begin{cases} 6x & \text{if } 0 \le x \le 40 \\ 9x - 120 & \text{if } x > 40 \end{cases}$$

Note that when an employee works more than 40 hours, she earns

$$\begin{array}{lrl} & 6(40) = & 240 \text{ for the first 40 hours} \\ \text{plus} & 9(x-40) = & \underline{9x-360} \text{ for the hours over 40} \\ & & 9x-120 \text{ for the total hours worked} \end{array}$$

EXAMPLE 7.2

Find the pay for an employee who works
(a) 36 hours
(b) 44 hours

SOLUTION 7.2

(a) Since 36 is less than 40, use $f(x) = 6x$. Then $f(x) = 6(36) =$ \$216.

(b) Since 44 is greater than 40, use $f(x) = 9x - 120$. Then

$f(x) = 9(44) - 120$	Substitute $x = 44$.
$= 396 - 120$	Multiply.
$= \$276$	Subtract.

Interval Notation

When we work with functions, we must often restrict our attention to certain sets of numbers (for example, we want profit to be greater than 0). We can use interval notation to describe these sets.

Interval Notation	Inequality	Graph
$(-1, 3)$	$-1 < x < 3$	 -1 3
$[-1, 3]$	$-1 \le x \le 3$	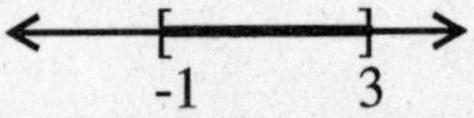 -1 3
$(-1, 3]$	$-1 < x \le 3$	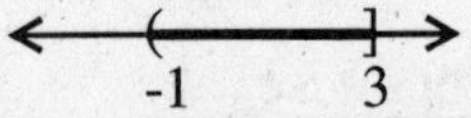 -1 3
$[-1, 3)$	$-1 \le x < 3$	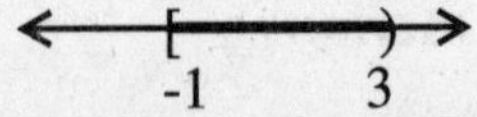 -1 3
$[2, \infty)$	$x \ge 2$	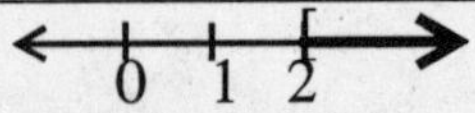 0 1 2
$(-\infty, 1)$	$x < 1$	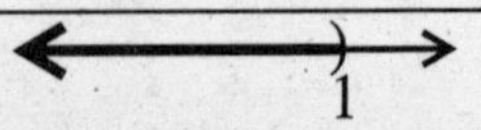 1

Notice that a parenthesis corresponds to an endpoint that is *not* included in the set (you may have graphed these as open dots in the past). A bracket corresponds to an endpoint that *is* included in the set (graphed as a closed dot in the past). To represent unbounded sets of numbers we use the symbols ∞ (positive infinity) and $-\infty$ (negative infinity).

EXAMPLE 7.3

Graph each interval on a number line.

(a) $(-2, 0)$

(b) $[1, 3]$

(c) $(-\sqrt{5}, 1]$

SOLUTION 7.3

(a) -2 0

(b) 1 3

(c) $-\sqrt{5}$ 1

EXAMPLE 7.4

Write each set in interval notation and then graph each set on a number line.

(a) $3 \le x \le 5$

(b) $-1 < x < 2$

(c) $-\sqrt{2} \le x < \sqrt{2}$

SOLUTION 7.4

(a) $[3, 5]$ Since the endpoints are included in the solution, use brackets.

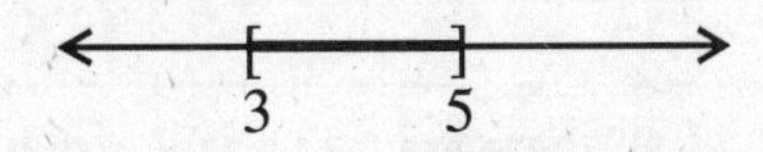

Endpoints on the graph match the interval notation.

(b) $(-1, 2)$ Since the endpoints are *not* included in the solution, use parentheses.

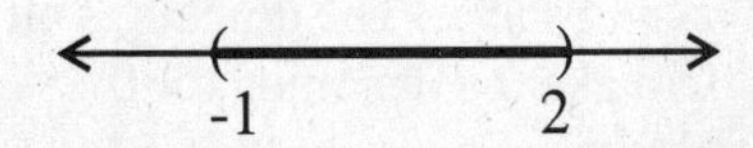

Endpoints on the graph match the interval notation.

(c) $[-\sqrt{2}, \sqrt{2})$ Since $x \ge -\sqrt{2}$, use a bracket on the left. Since $x < \sqrt{2}$, use a parenthesis on the right.

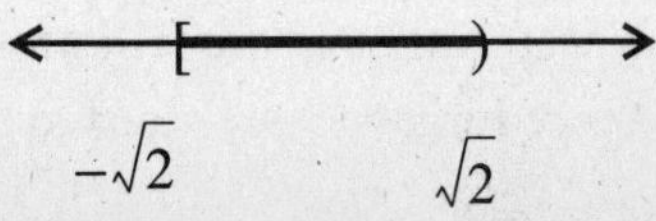

EXAMPLE 7.5

Complete the following table.

Interval Notation	Inequality	Graph
$[-1, 4)$		
	$x \geq -1$	
		←—→ 2

SOLUTION 7.5

Interval Notation	Inequality	Graph
$[-1, 4)$	$-1 \leq x < 4$	←[—)→ -1 4
$[-1, \infty)$	$x \geq -1$	←[—→ -1
$(-\infty, 2)$	$x < 2$	←—)→ 2

Domain and Range

If the domain of a function is *not* stated, it is assumed to be the largest set of real numbers which can be used as input values. Since we do not allow division by 0 or negative numbers under square root symbols (or *any* even-indexed roots), the following steps can be used to find the domain of a function when given an equation.

1. If there are no variables that give a value of 0 to the denominator or variables that have value less than 0 under a square root symbol, the domain is generally all real numbers.
2. If there is a variable in a denominator, set the denominator equal to 0 and solve. The domain is all real numbers *except* the values that make the denominator 0.
3. If there is a variable under a square root symbol, set the radicand greater than or equal to 0, and solve. The solution to the inequality is the domain.

Note: Although we will most often encounter square roots, the same rules apply for any even-indexed root such as $\sqrt[4]{x}$, $\sqrt[6]{x}$, etc.

EXAMPLE 7.6

Find the domain.

(a) $y = 2x + 1$

(b) $y = \dfrac{4}{x-2}$

(c) $y = \sqrt{3x+2}$

(d) $y = \dfrac{\sqrt{x-2}}{x-3}$

SOLUTION 7.6

(a) $y = 2x + 1$

The domain is the set of all real numbers since there are no variables in the denominator and no square roots.

(b) $y = \dfrac{4}{x-2}$

$x - 2 = 0$	Set the denominator equal to 0.
$x = 2$	Solve by adding 2 to both sides.

The domain is the set of all real numbers except 2, which we write as $(-\infty, 2) \cup (2, \infty)$.

(c) $y = \sqrt{3x+2}$

$3x + 2 \geq 0$	Set the radicand greater than or equal to 0.
$3x \geq -2$	Subtract 2 from both sides.
$x \geq -\dfrac{2}{3}$	Divide by 3.

The domain is the set of all real numbers greater than or equal to $-\dfrac{2}{3}$, written $\left[-\dfrac{2}{3}, \infty\right)$.

(d) $y = \dfrac{\sqrt{x-2}}{x-3}$

$x - 2 \geq 0$	Set the radicand greater than or equal to 0.
$x \geq 2$	Add 2 to both sides.

$x - 3 = 0$	Set the denominator equal to 0.
$x = 3$	Solve by adding 3 to both sides.

The domain must contain x values that are greater than or equal to 2 *and not* equal to 3. Therefore, the domain is $[2, 3) \cup (3, \infty)$.

Graphing Functions

The graph of a function provides a visual representation of input and output. These graphs can be derived from tables of values or by using a calculator or computer graphics. Note that the number of points needed to draw a graph varies. Choose as many x values as you need to sketch the function.

EXAMPLE 7.7

Graph each function.

(a) $f(x) = 2x$

(b) $g(x) = x^2 - 4x + 4$

(c) $f(x) = \begin{cases} 6x & \text{if } 0 \le x \le 40 \\ 9x - 120 & \text{if } x > 40 \end{cases}$

SOLUTION 7.7

(a) Use a table of values to graph $f(x) = 2x$:

x	$2x$	$f(x)$
0	2(0)	0
1	2(1)	2
2	2(2)	4

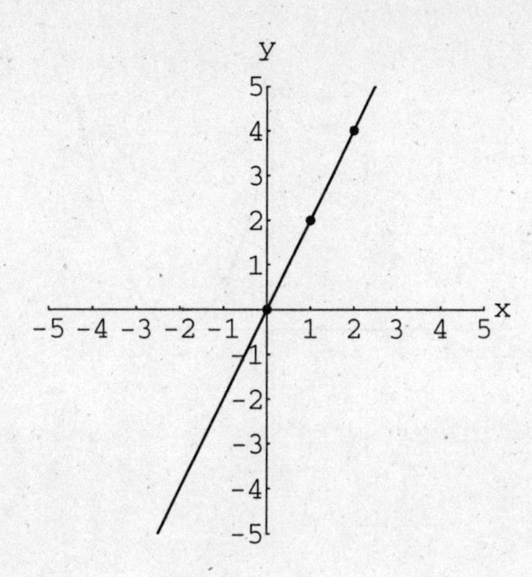

(b) $g(x) = x^2 - 4x + 4$

Make a table of values.

x	$x^2 - 4x + 4$	$g(x)$
–1	$(-1)^2 - 4(-1) + 4$	9
0	$(0)^2 - 4(0) + 4$	4
1	$(1)^2 - 4(1) + 4$	1
2	$(2)^2 - 4(2) + 4$	0
3	$(3)^2 - 4(3) + 4$	1

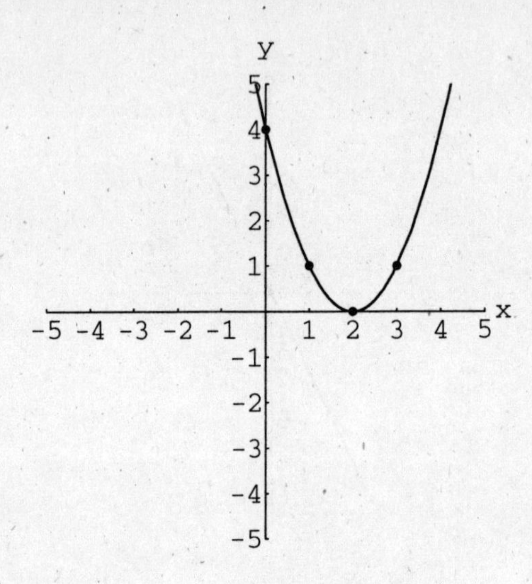

(c) $f(x) = \begin{cases} 6x & \text{if } 0 \leq x \leq 40 \\ 9x - 120 & \text{if } x > 40 \end{cases}$

We'll use two tables, one for $0 \leq x \leq 40$ and one for $x > 40$:

x	$6x$	$f(x)$
0	6(0)	0
10	6(10)	60
40	6(40)	240

x	$9x - 120$	$f(x)$
41	9(41) – 120	249
50	9(50) – 120	330

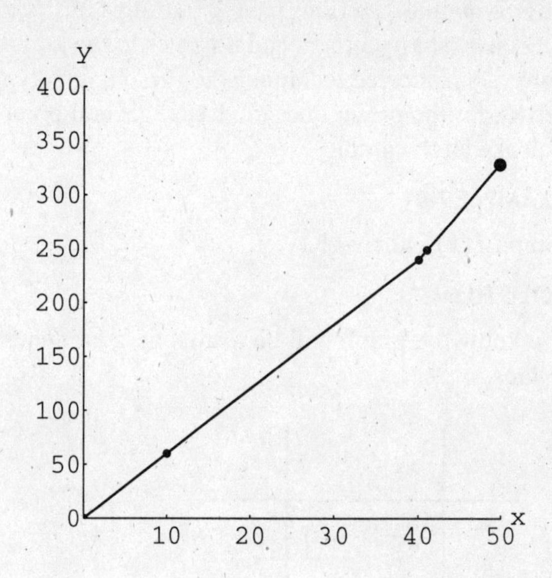

7.2 POLYNOMIAL AND RATIONAL FUNCTIONS

Although any function can be graphed by setting up a table of values and plotting points, it is helpful to recognize certain types of functions and their special features. In this section we will examine polynomial and rational functions.

Polynomial Functions

Polynomial functions contain sums and differences of terms where all variables have exponents that are nonnegative integers (i.e., x^2 and $x^3 + 4x^2$). The largest exponent is the **degree** of the polynomial function. For example,

$$f(x) = x^3 - 4x^2 + 2x - 1 \text{ and } g(x) = 5 - 8x$$

are polynomial functions of degree 3 and 1, respectively. The functions

$$h(x) = 4 - \sqrt{x} \text{ and } s(x) = \frac{4}{x} - \frac{6}{x^2}$$

are *not* polynomial functions because in each case there is an exponent present that is *not* a nonnegative integer ($\sqrt{x} = x^{1/2}$ and $4/x = 4x^{-1}$).

Polynomial functions have graphs that are smooth continuous curves—that is, no sharp corners and no holes in the graph. While we will develop some sophisticated techniques for graphing polynomial functions in chapter 10, for the present be sure to use several points (6 to 10 points) to help you sketch the graph.

EXAMPLE 7.8

Graph $f(x) = x^4 - 1$.

SOLUTION 7.8

We know the graph will be a smooth, continuous curve. Make a table of values.

x	$x^4 - 1$	$f(x)$
–3	$(-3)^4 - 1$	80
–2	$(-2)^4 - 1$	15
–1	$(-1)^4 - 1$	0
0	$(0)^4 - 1$	–1
1	$(1)^4 - 1$	0
2	$(2)^4 - 1$	15
3	$(3)^4 - 1$	80

Since the points (–3, 80) and (3, 80) have large y-values, you may choose not to show them on your graph. Or, you may change the scale on your graph paper by letting each block equal 10 units rather than 1 unit.

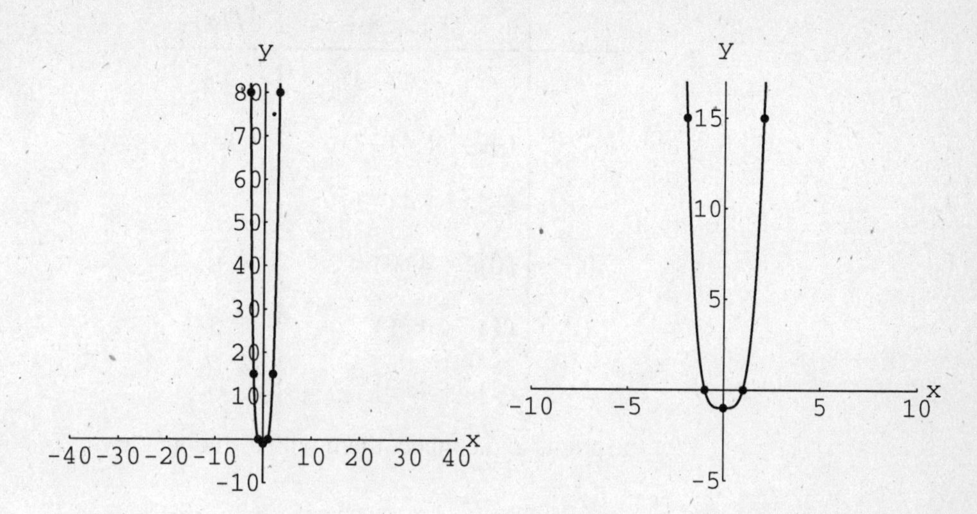

Graphs of $f(x) = x^4 - 1$ with Different Scales.

Quadratic Functions

A quadratic function is a polynomial function of degree 2. Quadratic functions can be written as $f(x) = ax^2 + bx + c$, (a, b, c constants) have graphs that are called **parabolas**, open upward if $a > 0$ and open downward if $a < 0$. The lowest point of a parabola that opens up (and the highest point of a parabola that opens down), is called the **vertex** of the parabola. In chapter 10 we will see how to find the vertex exactly. For now, graph each quadratic function using a table of values and approximate the vertex from your graph.

EXAMPLE 7.9

Graph each function.

(a) $f(x) = x^2 + 4x$

(b) $f(x) = -x^2 - 2x - 3$

SOLUTION 7.9

(a) $f(x) = x^2 + 4x = 1x^2 + 4x$. Since $a = 1 > 0$, the parabola opens up.

Make a table of values.

x	$x^2 + 4x$	$f(x)$
–4	$(-4)^2 + 4(-4)$	0

x	$x^2 + 4x$	$f(x)$
-3	$(-3)^2 + 4(-3)$	-3
-2	$(-2)^2 + 4(-2)$	-4
-1	$(-1)^2 + 4(-1)$	-3
0	$(0)^2 + 4(0)$	0
1	$(1)^2 + 4(1)$	5
2	$(2)^2 + 4(2)$	12

Plot the points and connect them with a smooth curve.

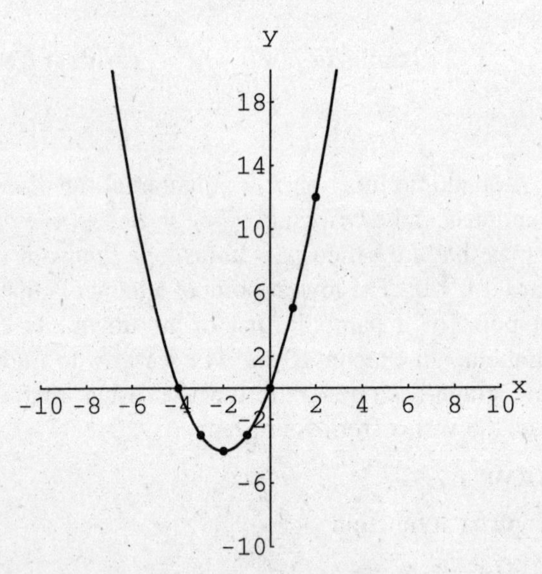

The vertex appears to be at (–2, –4).

(b) $f(x) = -x^2 - 2x - 3 = -1x^2 - 2x - 3$. Since $a = -1 < 0$, the parabola opens down. Make a table of values.

x	$-x^2 - 2x - 3$	$f(x)$
-3	$-(-3)^2 - 2(-3) - 3$	-6
-2	$-(-2)^2 - 2(-2) - 3$	-3
-1	$-(-1)^2 - 2(-1) - 3$	-2

x	$-x^2 - 2x - 3$	$f(x)$
0	$-(0)^2 - 2(0) - 3$	−3
1	$-(1)^2 - 2(1) - 3$	−6
2	$-(2)^2 - 2(2) - 3$	−11

Plot the points and connect them with a smooth curve.

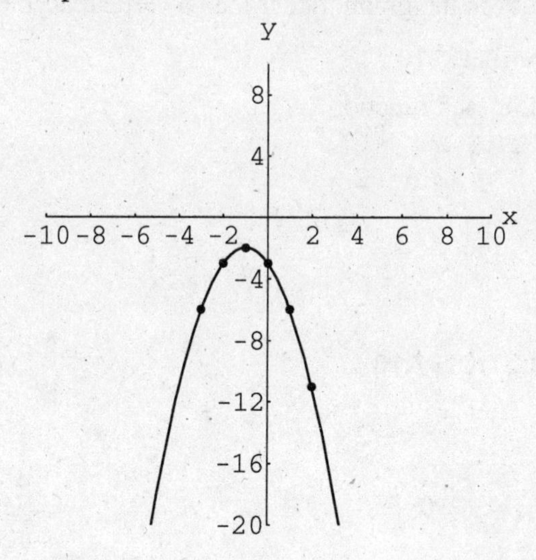

Rational Functions

A rational function can be written as a fraction, where the numerator and denominator are both polynomial functions. For example,

$$f(x) = \frac{4}{x-2} \text{ and } g(x) = \frac{x-1}{x^2+2x-3}$$

are rational functions. Recall from section 7.1 that we do not include x-values in the domain that would cause division by 0. Thus the domain of $f(x) = 4/(x-2)$ would be $(-\infty, 2) \cup (2, \infty)$. The values where the denominator of a rational function in reduced form equal 0 cause the graph to approach a **vertical asymptote**, a line that the curve gets closer and closer to but never crosses. (A rational function is in reduced form when the numerator and denominator contain no common factors.) Some steps to help you graph rational functions are:

1. Set the denominator equal to 0 and solve. You may need to factor.
2. Dot in vertical lines at each value found in step 1. (See note below for an exception to this.)

3. Make a table of values. Be sure to include values to the left **and** right of each vertical asymptote.
4. Connect points between the vertical asymptotes with a smooth curve. *Do not connect points across vertical asymptotes.*

Note: If a factor in the denominator is a factor of the numerator, the value found by setting that factor equal to 0 is excluded from the domain and the function at that x value is not plotted. We will describe this as leaving a "hole" in the graph, rather than a vertical asymptote.

EXAMPLE 7.10

Graph each function.

(a) $f(x) = \dfrac{4}{x-2}$

(b) $g(x) = \dfrac{x+1}{x^2+2x-3}$

SOLUTION 7.10

(a) $f(x) = \dfrac{4}{x-2}$

$x - 2 = 0$	Set the denominator equal to 0.
$x = 2$	Solve. Draw a dotted vertical line at $x = 2$.

Make a table of values. Include values to the left and right of 2.

x	$\dfrac{4}{x-2}$	$f(x)$
–1	$\dfrac{4}{(-1)-2}$	$-\dfrac{4}{3}$
0	$\dfrac{4}{(0)-2}$	–2
1	$\dfrac{4}{(1)-2}$	–4
2	$\dfrac{4}{(2)-2}$	undefined
3	$\dfrac{4}{(3)-2}$	4

x	$\frac{4}{x-2}$	$f(x)$
4	$\frac{4}{(4)-2}$	2
6	$\frac{4}{(6)-2}$	1

Plot the points. Connect the points on the left side and then connect the points on the right side of the vertical asymptote. *Do not connect points on the left to points on the right.*

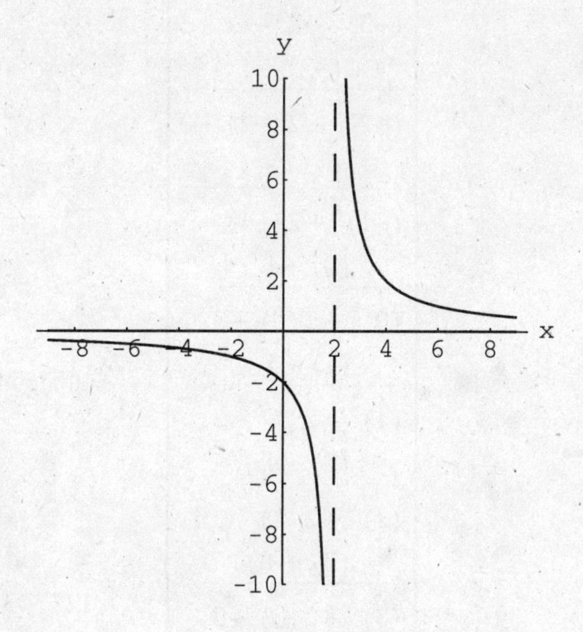

(b) $g(x) = \dfrac{x+1}{x^2+2x-3}$

$x^2 + 2x - 3 = 0$ — Set the denominator equal to 0.

$(x+3)(x-1) = 0$ — Factor to solve.

$x+3=0 \quad x-1=0$ — Set each factor equal to 0.

$x=-3 \quad x=1$ — Solve. Draw dotted vertical lines at $x=-3$ and $x=1$.

Male a table of values. Include values to the left and right of –3 and to the left and right of 1.

x	$\dfrac{x+1}{x^2+2x-3}$	$f(x)$
-5	$\dfrac{(-5)+1}{(-5)^2+2(-5)-3}$	$\dfrac{-4}{12} = -\dfrac{1}{3}$
-4	$\dfrac{(-4)+1}{(-4)^2+2(-4)-3}$	$\dfrac{-3}{5}$
-3	$\dfrac{(-3)+1}{(-3)^2+2(-3)-3}$	$\dfrac{-2}{0}$ = undefined
-2	$\dfrac{(-2)+1}{(-2)^2+2(-2)-3}$	$\dfrac{-1}{-3} = \dfrac{1}{3}$
-1	$\dfrac{(-1)+1}{(-1)^2+2(-1)-3}$	$\dfrac{0}{-4} = 0$
0	$\dfrac{(0)+1}{(0)^2+2(0)-3}$	$\dfrac{1}{-3} = -\dfrac{1}{3}$
1	$\dfrac{(1)+1}{(1)^2+2(1)-3}$	$\dfrac{2}{0}$ = undefined
2	$\dfrac{(2)+1}{(2)^2+2(2)-3}$	$\dfrac{3}{5}$
3	$\dfrac{(3)+1}{(3)^2+2(3)-3}$	$\dfrac{4}{12} = \dfrac{1}{3}$

Plot the points. Connect the points in each region defined by the vertical asymptotes.

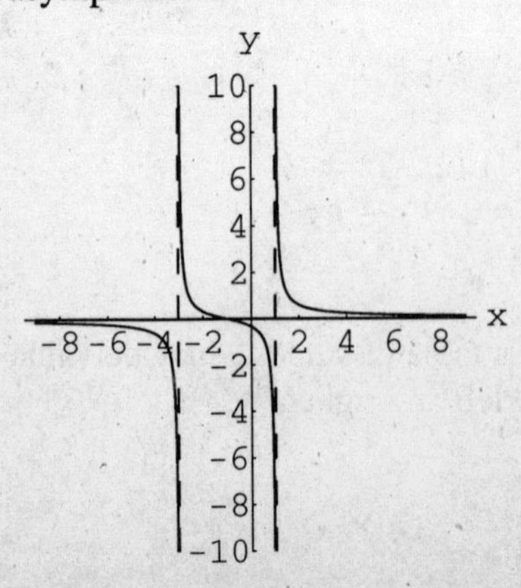

7.3 EXPONENTIAL FUNCTIONS

Graphing Exponential Functions

The functions studied previously had variables for the bases and nonnegative integers for exponents. When the exponent contains a variable and the base is positive and not equal to 1, e.g., $f(x) = 2^x$ we call the function an **exponential function**. Exponential functions have graphs that are smooth continuous curves (like the polynomial functions) but also have a **horizontal asymptote**. An exponential curve approaches its horizontal asymptote, but never crosses it.

EXAMPLE 7.11

Graph each function.

(a) $f(x) = 3^x$

(b) $f(x) = \left(\frac{1}{3}\right)^x$

SOLUTION 7.11

(a) $f(x) = 3^x$

Make a table of values:

x	3^x	$f(x)$
-3	3^{-3}	$\frac{1}{27}$
-2	3^{-2}	$\frac{1}{9}$
-1	3^{-1}	$\frac{1}{3}$
0	3^0	1
1	3^1	3
2	3^2	9
3	3^3	27

Plot the points and connect them with a smooth curve.

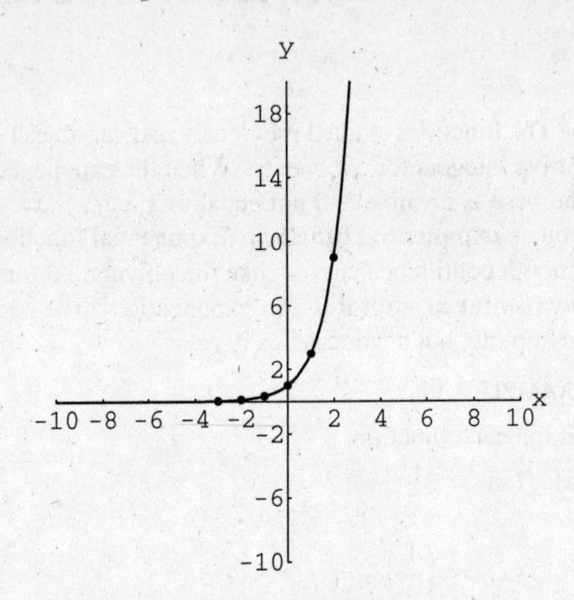

Note that the line $y = 0$ is a horizontal asymptote for this graph. The curve will approach this line as the x values get larger in a negative direction, but will never cross the asymptote.

(b) $f(x) = \left(\frac{1}{3}\right)^x$

Make a table of values:

x	$\left(\frac{1}{3}\right)^x$	$f(x)$
-3	$\left(\frac{1}{3}\right)^{-3}$	27
-2	$\left(\frac{1}{3}\right)^{-2}$	9
-1	$\left(\frac{1}{3}\right)^{-1}$	3
0	$\left(\frac{1}{3}\right)^{0}$	1
1	$\left(\frac{1}{3}\right)^{1}$	$\frac{1}{3}$

x	$\left(\frac{1}{3}\right)^x$	$f(x)$
2	$\left(\frac{1}{3}\right)^2$	$\frac{1}{9}$
3	$\left(\frac{1}{3}\right)^3$	$\frac{1}{27}$

Plot the points and connect them with a smooth curve.

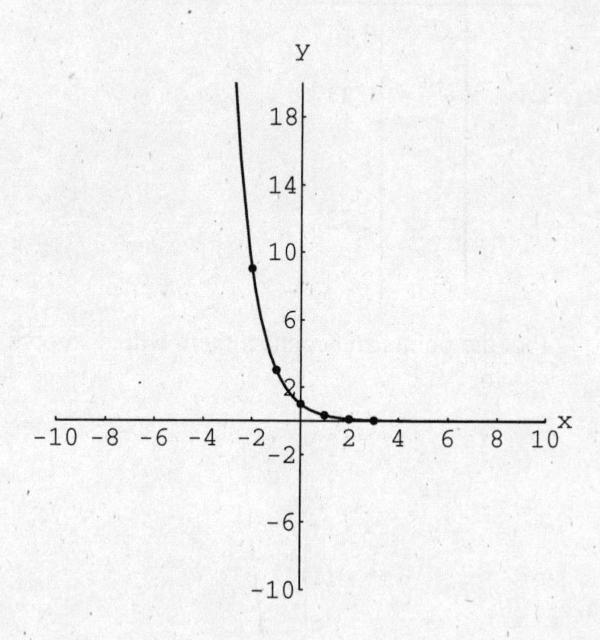

Base e

Applications that involve continuous compounding or growth or decay use a base of e, where e is the irrational number approximately equal to 2.7182818. We can find values of e^x using a calculator; usually, e^x is obtained by pressing the desired exponent, the SHIFT or INV or 2nd key, and then the ln key. For example, to find e^3 press

3
SHIFT or INV or 2nd
ln
= or EXE

Your display should show 20.08553. The number of digits displayed will depend on your particular brand of calculator.

We can graph exponential functions that have a base of e by making a table of values, plotting the points, and then connecting the points with a smooth curve.

EXAMPLE 7.12

Graph $f(x) = e^x$.

SOLUTION 7.12

Using a calculator, make a table of values. The values presented in our table have been rounded to two decimal places.

x	e^x	$f(x)$
–2	e^{-2}	0.14
–1	e^{-1}	0.37
0	e^0	1
1	e^1	2.72
2	e^2	7.39

Plot the points and connect them with a smooth curve.

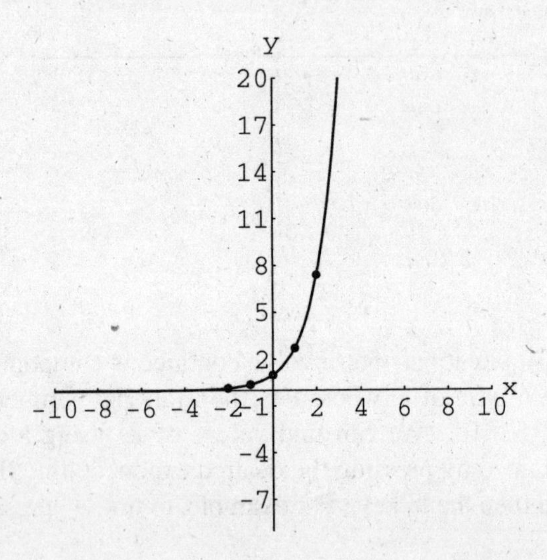

Applications

We will consider the growth and decay model

$$P = P_0 e^{rt}$$

where P_0 is the initial quantity at time $t = 0$, P is the quantity at time t, and r is the rate of growth or decay. (Note $r > 0$ for growth and $r < 0$ for decay.)

EXAMPLE 7.13

The population of a particular city at time t years is given by

$$P = P_0 e^{0.02t}$$

where the initial population in 1970 was 8000 people. Find the population in 1990.

SOLUTION 7.13

Since the initial population was 8000, we know $P_0 = 8000$. We are looking for the population 20 years after the initial population was counted, so $t = 20$.

$P = P_0 e^{0.02t}$	Write the equation.
$P = 8000e^{0.02(20)}$	Substitute $P_0 = 8000$ and $t = 20$.
$P = 8000e^{0.4}$	Multiply.
$P \approx 8000(1.4918)$	Use a calculator to find $e^{0.4}$.
$\approx 11,934$	The answer is the approximate population in 1990.

7.4 LOGARITHMIC FUNCTIONS

In section 7.3 we discussed exponential functions. Logarithms are the inverse functions of exponential functions. (Note: f and g are inverse functions if $f(g(x)) = g(f(x)) = x$.)

> **Definition**. $\log_a x = y$ is defined by requiring $a^y = x$ for $a > 0$, $a \neq 1$.

$\log_a x$ is read "the log of x to the base a," or "log base a of x." A visual aid using a small loop may help you translate from logarithmic form to exponential form:

$$\log_2 8 = 3 \quad \text{means} \quad 2^3 = 8$$

$$\log_4 16 = 2 \quad \text{means} \quad 4^2 = 16$$

$$\log_{10} 1000 = 3 \quad \text{means} \quad 10^3 = 1000$$

EXAMPLE 7.14

Rewrite each expression in exponential form.

(a) $\log_6 36 = 2$

(b) $\log_{10} 0.01 = -2$

(c) $\log_5 1 = 0$

(d) $\log_8 4 = \frac{2}{3}$

SOLUTION 7.14

(a) $\log_6 36 = 2$ means $6^2 = 36$

(b) $\log_{10} 0.01 = -2$ means $10^{-2} = 0.01$

(c) $\log_5 1 = 0$ means $5^0 = 1$

(d) $\log_8 4 = \frac{2}{3}$ means $8^{2/3} = 4$

EXAMPLE 7.15

Rewrite each expression in logarithmic form.

(a) $10^2 = 100$

(b) $4^{-3} = \frac{1}{64}$

(c) $16^{3/4} = 8$

SOLUTION 7.15

(a) $10^2 = 100$ means $\log_{10} 100 = 2$

The logarithm is the exponent, so 2 is the logarithm.

(b) $4^{-3} = \frac{1}{64}$ means $\log_4 \frac{1}{64} = -3$

(c) $16^{3/4} = 8$ means $\log_{16} 8 = \frac{3}{4}$

Properties of Logarithms

Several properties of logarithms can be used to simplify calculations involving logarithms.

Properties of Logarithms

1. $\log_a(xy) = \log_a x + \log_a y$
2. $\log_a\left(\frac{x}{y}\right) = \log_a x - \log_a y$
3. $\log_a x^b = b \log_a x$
4. $\log_a 1 = 0$
5. $\log_a(a^x) = x$
6. $a^{\log_a x} = x$

EXAMPLE 7.16

Use the properties of logarithms to simplify each expression.

(a) $\log_4 5 + \log_4 3$

(b) $\log_7 45 - \log_7 9$

(c) $\log_2 x^3$

(d) $3^{\log_3 8}$

SOLUTION 7.16

(a) $\log_4 5 + \log_4 3$

$= \log_4 (5)(3)$ Use property 1.

$= \log_4 15$ Simplify.

(b) $\log_7 45 - \log_7 9$

$= \log_7\left(\frac{45}{9}\right)$ Use property 2.

$= \log_7 5$ Simplify.

(c) $\log_2 x^3$

$= 3\log_2 x$ Use property 3.

(d) $3^{\log_3 8}$

$= 8$ Use property 6.

EXAMPLE 7.17

Rewrite each expression as a sum, difference, or product of logarithms.

(a) $\log_2 5x$

(b) $\log_{10}\left(\frac{3m}{n}\right)$

(c) $\log_5 \sqrt{x}$

SOLUTION 7.17

(a) $\log_2 5x = \log_2 5 + \log_2 x$ Since $5x$ is a product, use property 1.

(b) $\log_{10}\left(\frac{3m}{n}\right) = \log_{10} 3m - \log_{10} n$ Use property 2 to rewrite the quotient.

$= \log_{10} 3 + \log_{10} m - \log_{10} n$ Use property 1 to rewrite the log of the product $3m$.

(c) $\log_5 \sqrt{x} = \log_5 x^{1/2}$ Rewrite $\sqrt{x} = x^{1/2}$.

$= \frac{1}{2}\log_5 x$ Use property 3 to write the exponent in front of the log.

EXAMPLE 7.18

Rewrite each expression as a sum, difference, or product of logarithms.

(a) $\log_3\left(\frac{4xy}{pq}\right)$

(b) $\log_5 x^2 y^3$

(c) $\log_2\left(\frac{x\sqrt[3]{y}}{z^5}\right)$

SOLUTION 7.18

(a) $\log_3\left(\frac{4xy}{pq}\right)$

$= \log_3 4xy - \log_3 pq$ Use property 2 to rewrite the log of a quotient.

$= \log_3 4 + \log_3 x + \log_3 y$ — Use property 1 to rewrite the log of each product.

$- (\log_3 p + \log_3 q)$

$= \log_3 4 + \log_3 x + \log_3 y$ — Distribute the negative to remove the parentheses.

$- \log_3 p - \log_3 q$

(b) $\log_5 x^2 y^3$

$= \log_5 x^2 + \log_5 y^3$ — Use property 1.

$= 2\log_5 x + 3\log_5 y$ — Use property 3.

(c) $\log_2 \left(\frac{x\sqrt[3]{y}}{z^5} \right)$

$= \log_2 x\sqrt[3]{y} - \log_2 z^5$ — Use property 2.

$= \log_2 xy^{1/3} - \log_2 z^5$ — Rewrite $\sqrt[3]{y} = y^{1/3}$.

$= \log_2 x + \log_2 y^{1/3} - \log_2 z^5$ — Use property 1.

$= \log_2 x + \frac{1}{3}\log_2 y - 5\log_2 z$ — Use property 3.

Graphing Logarithmic Functions

It is usually easier to rewrite a logarithmic function in exponential form before setting up a table of values to graph a logarithmic function. The steps we'll use to graph logarithmic functions are:

1. Rewrite the logarithmic equation in exponential form.
2. Make a table of values, choosing y-values, and finding x-values.
3. Plot the points listed in the table and connect them with a smooth curve.

EXAMPLE 7.19

Graph each logarithmic function.

(a) $y = \log_2 x$

(b) $y = \log_{1/2} x$

SOLUTION 7.19

(a) $y = \log_2 x$ means $2^y = x$

Rewrite the given equation in exponential form.

Make a table, choosing y-values and finding x-values. Be careful to plot points as (x, y) even though the table has been set up for convenience with the y's listed first.

y	2^y	x
-2	2^{-2}	$\frac{1}{4}$
-1	2^{-1}	$\frac{1}{2}$
0	2^0	1
1	2^1	2
2	2^2	4

Plot the points and connect them with a smooth curve.

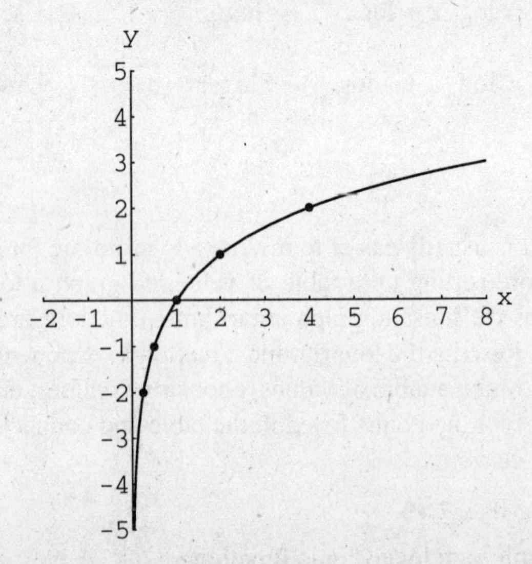

(b) $y = \log_{1/2} x$ means $\left(\frac{1}{2}\right)^y = x$

Rewrite the given equation in exponential form.

Make a table, choosing y-values and finding x-values.

y	$\left(\frac{1}{2}\right)^y$	x
-2	$\left(\frac{1}{2}\right)^{-2}$	4
-1	$\left(\frac{1}{2}\right)^{-1}$	2
0	$\left(\frac{1}{2}\right)^{0}$	1
1	$\left(\frac{1}{2}\right)^{1}$	$\frac{1}{2}$
2	$\left(\frac{1}{2}\right)^{2}$	$\frac{1}{4}$

Plot the points and connect them with a smooth curve.

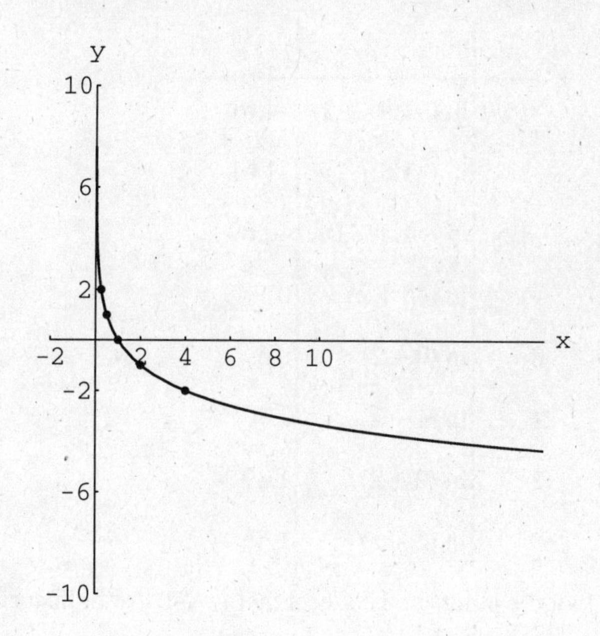

ln x

The **natural logarithm of *x***, written **ln *x***, is a logarithm with a base of e. That is,

$$\log_e x = \ln x$$

We can find values of ln x using a calculator. Usually, ln x is obtained by pressing x then the ln key. For example, to find ln 5, press

5
ln
= or EXE
Your display should show 1.609437912. The number of digits displayed will depend on your particular brand of calculator.

Graphing ln x

We can graph natural logarithmic functions by making a table of values. Since we can use a calculator to find the output values, we do *not* need to first rewrite the function in exponential form.

EXAMPLE 7.20

Graph $f(x) = \ln(x+2)$.

SOLUTION 7.20

Make a table of values. The values presented in our table have been rounded to two decimal places.

x	$\ln(x+2)$	$f(x)$
–1.99	$\ln(-1.99+2)$	–4.61
–1.8	$\ln(-1.8+2)$	–1.61
–1.5	$\ln(-1.5+2)$	–0.69
–1	$\ln(-1+2)$	0
0	$\ln(0+2)$	0.69
1	$\ln(1+2)$	1.10
2	$\ln(2+2)$	1.39
3	$\ln(3+2)$	1.61

Plot the points and connect them with a smooth curve.

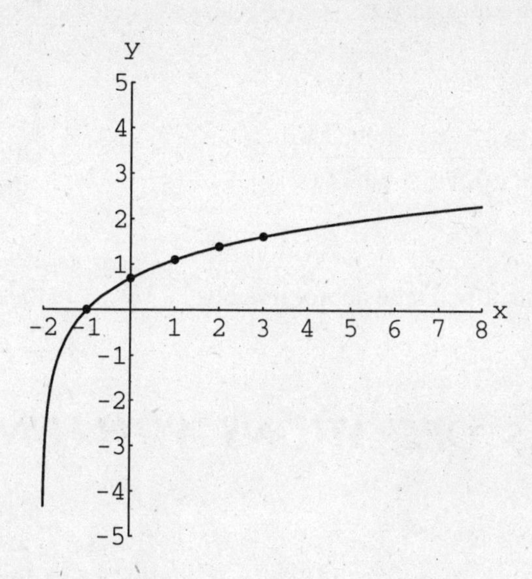

Applications

In section 4.2 we used the compound interest formula to find the amount of money accrued after a given amount of time. We can use logarithms to work with the compound interest formula to find the length of time needed to accrue a certain amount of money.

EXAMPLE 7.21

The Smiths wish to save \$40,000 for college tuition for their 1-year-old daughter. If they invest \$10,000 in an account paying 8% interest compounded quarterly, how long will it take for the account to be worth \$40,000?

SOLUTION 7.21

$$A = P\left(1 + \frac{r}{n}\right)^{nt}$$ Use the compound interest formula.

$$40{,}000 = 10{,}000\left(1 + \frac{0.08}{4}\right)^{4t}$$ Substitute the given values.

$$40{,}000 = 10000(1.02)^{4t}$$ Simplify inside the parentheses.

$$\frac{40000}{10000} = \frac{10000\,(1.02)^{4t}}{10000}$$ Divide both sides by 10,000.

$$4 = (1.02)^{4t}$$ Simplify both sides.

$$\ln 4 = \ln\,(1.02)^{4t}$$ Take the natural logarithm of both sides.

$\ln 4 = 4t \ln (1.02)$ — Use property 3 of logarithms to write the exponent in front of ln.

$$\frac{\ln 4}{4\ln(1.02)} = \frac{4t\ln(1.02)}{4\ln(1.02)}$$ — Divide both sides by 4 ln (1.02).

$17.5 \approx t$ — Use a calculator to find the value of the left side.

Thus it will take approximately 17 1/2 years to save $40,000.

7.5 OPERATIONS WITH FUNCTIONS

Operations with Functions

Just as we can add, subtract, multiply and divide numbers, we can add, subtract, multiply and divide functions.

Definitions.

If f and g are functions,

$(f+g)(x) = f(x) + g(x)$

$(f-g)(x) = f(x) - g(x)$

$(fg)(x) = f(x)g(x)$

$$\frac{f}{g}(x) = \frac{f(x)}{g(x)}, \; g(x) \neq 0$$

EXAMPLE 7.22

If $f(x) = 2x + 5$ and $g(x) = x^2 - 2x + 1$, find

(a) $(f+g)(3)$

(b) $(fg)(-1)$

(c) $\frac{f}{g}(0)$

SOLUTION 7.22

(a) $(f+g)(3)$

$= f(3) + g(3)$ — Use $(f+g)(x) = f(x) + g(x)$.

$= 2(3) + 5 + (3)^2 - 2(3) + 1$ — Substitute $x = 3$.

$= 6 + 5 + 9 - 6 + 1$ — Simplify.

$= 15$

(b) $(fg)(-1)$

$= f(-1)g(-1)$ — Use $(fg)(x) = f(x)g(x)$.

$= [2(-1) + 5][(-1)^2 - 2(-1) + 1]$ — Substitute $x = -1$.

$= (3)(4)$ — Simplify inside each set of brackets.

$= 12$

(c) $\frac{f}{g}(0) = \frac{f(0)}{g(0)}$ — Use $\frac{f}{g}(x) = \frac{f(x)}{g(x)}$

$= \frac{2(0) + 5}{(0)^2 - 2(0) + 1}$ — Substitute $x = 0$.

$= \frac{5}{1} = 5$ — Simplify.

EXAMPLE 7.23

If $f(x) = 3x^2 - 5x - 4$ and $g(x) = 2x + 3$, find

(a) $(f - g)(x)$

(b) $(fg)(x)$

(c) $\frac{f}{g}(x)$

SOLUTION 7.23

(a) $(f - g)(x)$

$= f(x) - g(x)$ — Use the definition of function subtraction.

$= (3x^2 - 5x - 4) - (2x + 3)$ — Substitute.

$= 3x^2 - 5x - 4 - 2x - 3$ — Distribute.

$= 3x^2 - 7x - 7$ — Combine similar terms.

(b) $(fg)(x)$

$= f(x)g(x)$ — Use the definition of function multiplication.

$= (3x^2 - 5x - 4)(2x + 3)$ — Substitute.

$= 6x^3 + 9x^2 - 10x^2 - 15x - 8x - 12$ — Distribute.

$= 6x^3 - x^2 - 23x - 12$ Combine similar terms.

(c) $\frac{f}{g}(x) = \frac{f(x)}{g(x)}$ Use the definition of function division.

$= \frac{3x^2 - 5x - 4}{2x + 3}$ Substitute.

Composition of Functions

In addition to the four basic operations (addition, subtraction, multiplication, division), we can also form composite functions.

> **Definition.**
> The composition of f and g, written $f(g(x))$ means to evaluate f at $g(x)$.

The composition of functions indicates an order of evaluating the functions. $f(g(3))$ means first evaluate $g(3)$, then evaluate f at the resulting value.

EXAMPLE 7.24

If $f(x) = 3x + 1$ and $g(x) = -2x - 4$, find $f(g(3))$.

SOLUTION 7.24

$f(g(3))$ Use the definition of composition of functions.

$g(3) = -2(3) - 4$ First find $g(3)$.

$= -10$ Now evaluate $f(-10)$.

$f(g(3)) = f(-10)$

$= 3(-10) + 1$ Substitute $x = -10$.

$= -29$

EXAMPLE 7.25

If $f(x) = 3x + 1$ and $g(x) = -2x - 4$, find $g(f(3))$.

SOLUTION 7.25

$f(3) = 3(3) + 1 = 10$ First find $f(3)$.

$g(f(3)) = g(10)$ Now evaluate $g(10)$.

$= -2(10) - 4$ Substitute $x = 10$.

$= -24$

Notice from examples 7.24 and 7.25 that in general, $f(g(x)) \neq g(f(x))$.

EXAMPLE 7.26

If $f(x) = x^2 - 1$ and $g(x) = 3x + 4$, find $f(g(x))$ and $g(f(x))$.

SOLUTION 7.26

Find $f(g(x))$.
We know $g(x) = 3x + 4$, so we must find $f(3x + 4)$.
$f(3x + 4)$

$= (3x + 4)^2 - 1$	Replace x with $3x + 4$.
$= 9x^2 + 24x + 16 - 1$	Multiply.
$= 9x^2 + 24x + 15$	Add similar terms.

Find $g(f(x))$
We know $f(x) = x^2 - 1$, so we must find $g(x^2 - 1)$.

$g(x^2 - 1) =$

$= 3(x^2 - 1) + 4$	Replace x with $x^2 - 1$.
$= 3x^2 - 3 + 4$	Multiply.
$= 3x^2 + 1$	Add similar terms.

Practice Exercises

1. If $g(x) = 2x^2 - 3x + 1$ find

   (a) $g(0)$

   (b) $g(-2)$

   (c) $g(x+h)$

2. Find the domain for each function. Write the answers using interval notation.

   (a) $y = 3x - 2$

   (b) $y = \dfrac{6}{x+1}$

   (c) $y = \sqrt{3x-4}$

   (d) $y = \dfrac{x+1}{\sqrt{x-3}}$

3. Graph each function.

   (a) $f(x) = 3x + 1$

   (b) $f(x) = x^3 + 2$

   (c) $f(x) = -x^2 + 2x - 3$

   (d) $f(x) = \dfrac{x-1}{x^2 - x - 2}$

   (e) $f(x) = 4^x$

4. Rewrite each expression in exponential form.

   (a) $\log_5 25 = 2$

   (b) $\log_6 1 = 0$

5. Rewrite each expression in logarithmic form.

   (a) $7^2 = 49$

   (b) $8^{2/3} = 4$

6. Use the properties of logarithms to simplify each expression.

   (a) $\log_2 3 + \log_2 4$

   (b) $\log_5 15 - \log_5 3$

   (c) $\log_4 (4^5)$

7. Rewrite each expression as a sum, difference or product of logarithms.

   (a) $\log_2 \left(\dfrac{5x}{y^2}\right)$

   (b) $\log_2 \left(\dfrac{\sqrt{xy}}{z^2}\right)$

8. If $f(x) = 2x^2 - 3x$ and $g(x) = 4x - 1$, find

   (a) $(f+g)(x)$

   (b) $(fg)(x)$

   (c) $f(g(2))$

   (d) $g(f(x))$

Answers

1. (a) 1

   (b) 15

   (c) $2x^2 + 4xh + 2h^2 - 3x - 3h + 1$

2. (a) $(-\infty, \infty)$

   (b) $(-\infty, -1) \cup (-1, \infty)$

   (c) $[4/3, -\infty)$

   (d) $(3, \infty)$

3. (a)

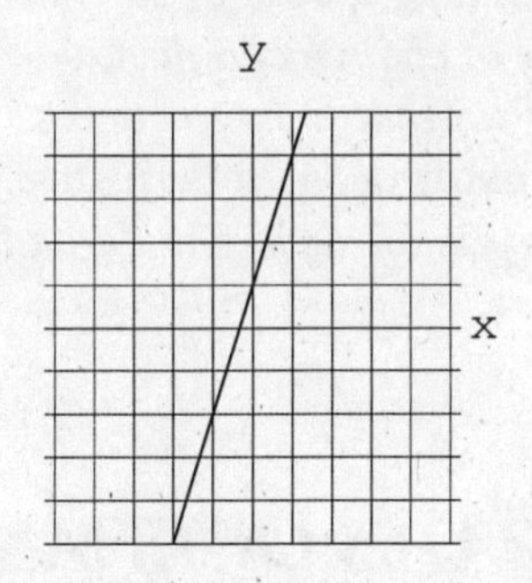

   (b)

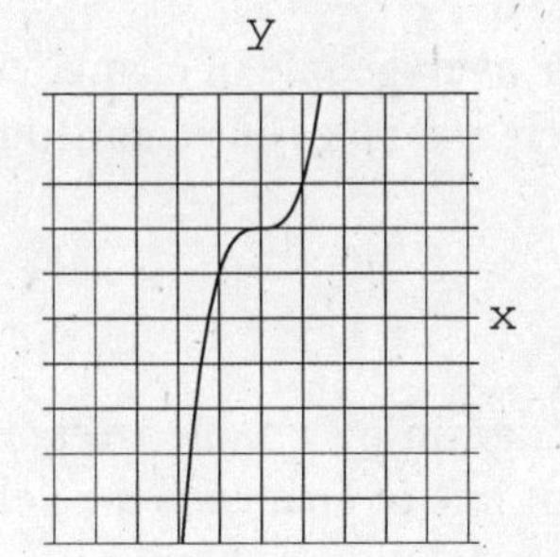

   (c)

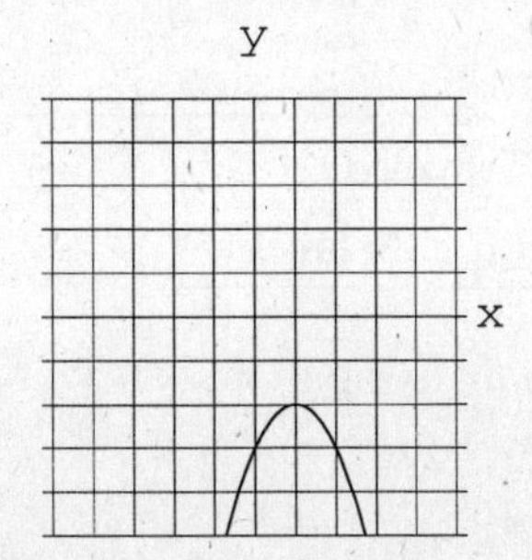

   (d)

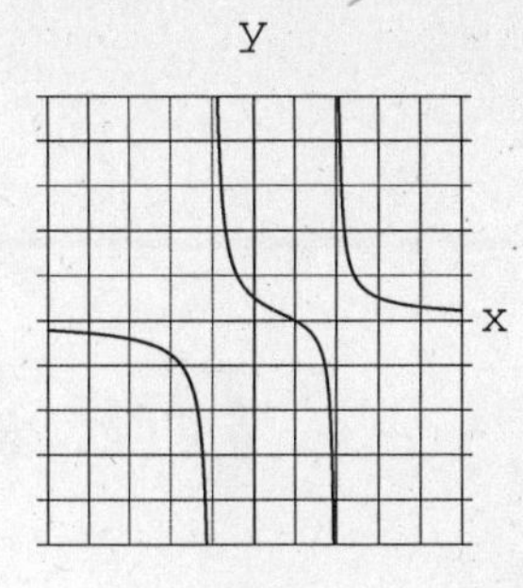

   (e)

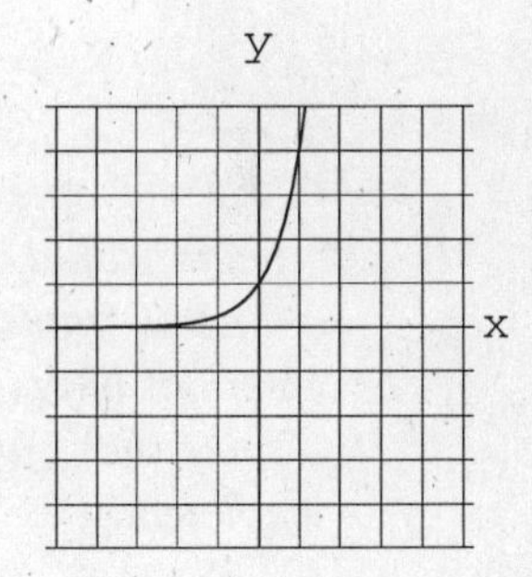

4. (a) $5^2 = 25$

   (b) $6^0 = 1$

5. (a) $\log_7 49 = 2$

   (b) $\log_8 4 = 2/3$

6. (a) $\log_2 12$

   (b) 1

   (c) 5

7. (a) $\log_2 5 + \log_2 x - 2\log_2 y$

   (b) $\frac{1}{2}\log_2 x + \frac{1}{2}\log_2 y - 2\log_2 z$

8. (a) $2x^2 + x - 1$

   (b) $8x^3 - 14x^2 + 3x$

   (c) 77

   (d) $8x^2 - 12x - 1$

8

The Derivative

This chapter introduces one of the building blocks of calculus. We will begin with a discussion of average rate of change and then look at instantaneous rates of change and slopes of tangent lines to curves. The concept of limits will be introduced to formally define a derivative, and then some shortcut rules will be presented to enable us to find derivatives more efficiently.

8.1 INSTANTANEOUS RATES OF CHANGE

This section begins by discussing the average rate of change of distance over a certain amount of time and then examines instantaneous rates of change.

Average Rate of Change

A rate of change measures the change in a variable over a specified time interval. One common example of rate of change is the velocity of a moving object. We can define average velocity as follows.

$$\textbf{Average Velocity} = \frac{\text{Distance Traveled}}{\text{Time to Travel the Distance}}$$

A car that travels 120 miles from 2 p.m. to 4 p.m. has an average velocity of

$$\frac{120 \text{ miles}}{4 \text{ p.m.} - 2 \text{ p.m.}} = \frac{120 \text{ miles}}{2 \text{ hrs.}} = 60 \text{ miles per hour}$$

Table 8.1 lists the distance (in miles) two cars have traveled after leaving location *A*. The time at which these distances were reached is listed in the Time column.

	Distance from Location A	Time
car 1	10 miles 130 miles	9 a.m. 11 a.m.
car 2	30 miles 100 miles	10 a.m. 11 a.m.

Table 8.1 Time and Distance Table for Two Cars

EXAMPLE 8.1

Using the data provided in Table 8.1, find:
(a) the average velocity of car 1
(b) the average velocity of car 2

SOLUTION 8.1

(a) $\text{average velocity} = \dfrac{\text{Distance Traveled}}{\text{Time to Travel}}$ Use the average velocity formula.

total distance covered = 130 – 10 miles
total time = 11 a.m. – 9 a.m. Find the total distance covered and total time.

$\text{average velocity} = \dfrac{(130 - 10)\text{ miles}}{11\text{ a.m.} - 9\text{ a.m.}}$ Substitute into the formula.

$= \dfrac{120\text{ miles}}{2\text{ hours}}$

$= 60$ miles per hour Simplify.

(b) $\text{average velocity} = \dfrac{\text{Distance Traveled}}{\text{Time to Travel}}$ Use the average velocity formula.

total distance covered = 100 – 30 miles
total time = 10 a.m. – 9 a.m. Find the total distance covered and total time.

$\text{average velocity} = \dfrac{100 - 30\text{ miles}}{10\text{ a.m.} - 9\text{ a.m.}}$ Substitute into the formula.

$$= \frac{70 \text{ miles}}{1 \text{ hour}} \qquad \text{Simplify.}$$
$$= 70 \text{ miles per hour}$$

Instantaneous Rate of Change

If we let the time interval approach 0 in our average velocity formula, it seems reasonable to assume the ratio D/t should approach the instantaneous velocity—that is, the velocity at a given instant of time. Can you find a trend in the following table containing average velocities for a car that travels a distance of

$$s(t) = 6t^2 + 24t$$

miles in t hours?

Time Interval	Average Velocity
$t = 2$ to $t = 3$	$\frac{s(3) - s(2)}{3 - 2} = \frac{126 - 72}{1} = 54$
$t = 2$ to $t = 2.5$	$\frac{s(2.5) - s(2)}{2.5 - 2} = \frac{97.5 - 72}{0.5} = 51$
$t = 2$ to $t = 2.1$	$\frac{s(2.1) - s(2)}{2.1 - 2} = \frac{76.86 - 72}{0.1} = 48.6$
$t = 2$ to $t = 2.001$	$\frac{s(2.001) - s(2)}{2.001 - 2} = \frac{72.0488006 - 72}{0.001} =$ 48.006

Perhaps you noticed that the average velocities were approaching 48 mph as the time interval got smaller. Indeed the instantaneous velocity is 48 mph, which we will demonstrate more clearly in example 8.2 (c).

We can generalize our discussion of instantaneous velocity to include the rate of change of variables other than distance. We use a formula, called the **Difference Quotient**, and let $f(x)$ represent the function.

The Difference Quotient

$$\frac{f(x+h) - f(x)}{h}$$

The following rules will help you use the Difference Quotient.

Rules for Using the Difference Quotient

1. Find $f(x+h)$ by replacing x with $(x+h)$ in the given function. Simplify.
2. Substitute the simplified expression for $f(x+h)$ and $f(x)$ in the Difference Quotient formula.
3. Simplify.

If the function has an instantaneous rate of change then as h approaches 0, the Difference Quotient approaches the instantaneous rate of change of f at x.

EXAMPLE 8.2

Find the instantaneous rate of change for each function.

(a) $f(x) = x^2$

(b) $f(x) = 3x+1$

(c) $f(x) = 6x^2+24x$

(d) $f(x) = \frac{1}{x}$

SOLUTION 8.2

(a) $f(x) = x^2$ — Copy the given function.

$f(x+h) = (x+h)^2$ — Replace x with $(x+(h)$.

$= x^2+2xh+h^2$ — Simplify.

$$\frac{f(x+h)-f(x)}{h} = \frac{(x^2+2xh+h^2)-(x^2)}{h}$$

Substitute into the Difference Quotient.

$$= \frac{x^2+2xh+h^2-x^2}{h}$$

Simplify.

$$= \frac{2xh+h^2}{h}$$

Add similar terms.

$$= \frac{h(2x+h)}{h}$$

Factor out h.

$= 2x + h$ Reduce.

As h approaches 0, $2x + h$ approaches $2x$. Thus, the instantaneous rate of change of f at x is $2x$.

(b) $f(x) = 3x + 1$ Copy the given function.

$f(x+h) = 3(x+h) + 1$ Replace x with $(x + (h)$.

$= 3x + 3h + 1$ Simplify.

$$\frac{f(x+h) - f(x)}{h} = \frac{(3x + 3h + 1) - (3x + 1)}{h}$$

Substitute into the Difference Quotient.

$$= \frac{3x + 3h + 1 - 3x - 1}{h}$$

Distribute.

$$= \frac{3h}{h}$$

Add similar terms.

$= 3$ Reduce.

As h approaches 0, the Difference Quotient is 3. Thus, the instantaneous rate of change of f at x is 3.

(c) $f(x) = 6x^2 + 24x$ Copy the given function.

$f(x+h) = 6(x+h)^2 + 24(x+h)$ Replace x with $(x + (h)$.

$= 6(x^2 + 2xh + h^2) + 24x + 24h$

$= 6x^2 + 12xh + 6h^2 + 24x + 24h$

Simplify.

$$\frac{f(x+h) - f(x)}{h}$$

$$= \frac{(6x^2 + 12xh + 6h^2 + 24x + 24h) - (6x^2 + 24x)}{h}$$

Substitute into the Difference Quotient.

$$= \frac{6x^2 + 12xh + 6h^2 + 24x + 24h - 6x^2 - 24x}{h}$$

$$= \frac{12xh + 6h^2 + 24h}{h}$$ Add similar terms.

$$= \frac{h(12x + 6h + 24)}{h}$$ Factor out h.

$$= 12x + 6h + 24$$ Reduce.

As h approaches 0, the Difference Quotient approaches $12x + 24$. Therefore the instantaneous rate of change of f at x is $12x + 24$. Note that when $x = 2$, this means $12x + 24 = 24 + 24 = 48$, the same rate of change found in our previous discussion of instantaneous velocity.

(d) $$f(x) = \frac{1}{x}$$ Copy the given function.

$$f(x+h) = \frac{1}{x+h}$$ Replace x with $(x + (h)$.

$$\frac{f(x+h) - f(x)}{h} = \frac{\frac{1}{x+h} - \frac{1}{x}}{h}$$ Substitute for $f(x)$ and $f(x+h)$ in the Difference Quotient

$$\frac{\frac{x}{x} \cdot \frac{1}{x+h} - \frac{1}{x} \cdot \frac{x+h}{x+h}}{h}$$ Use the common denominator $x(x+h)$ to combine the terms in the numerator.

$$\frac{\frac{x - x - h}{x(x+h)}}{h}$$ Distribute.

$$\frac{\frac{-h}{x(x+h)}}{h}$$ Combine like terms in the numerator.

$$\frac{\frac{-h}{x(x+h)}}{\frac{h}{1}} = \frac{-h}{x(x+h)} \cdot \frac{1}{h}$$ Convert h to $h/1$ and invert the denominator to multiply.

$$\frac{-\cancel{h}}{x(x+h)} \cdot \frac{1}{\cancel{h}} = \frac{-1}{x(x+h)}$$ The h factors cancel to form the solution.

$$\frac{-1}{x(x+h)} = \frac{-1}{x^2 + hx}$$

As h approaches 0, the Difference Quotient approaches $-1/x^2$. Thus the instantaneous rate of change of f at x is $-1/x^2$. Note that there is *not* an instantaneous rate of change at $x = 0$ where $f(x)$ was undefined.

Slope of the Tangent Line

If we use $(x, f(x))$ as a point on the curve $y = f(x)$, and $(x+h, f(x+h))$ as a second point on that curve, the difference quotient represents the slope of the line through those two points.

$$m = \frac{y_2 - y_1}{x_2 - x_1} = \frac{f(x+h) - f(x)}{(x+h) - x} = \frac{f(x+h) - f(x)}{h}$$

If the limit exists as h approaches 0, this line becomes a **tangent line** to the curve.

EXAMPLE 8.3

Find the equation of the tangent line to $f(x) = x^2 - 2x + 5$ at (3, 8).

SOLUTION 8.3

Use the Difference Quotient to find the slope of the tangent line.

$f(x) = x^2 - 2x + 5$ — Copy the given function.

$f(x+h) = (x+h)^2 - 2(x+h) + 5$ — Replace x with $(x + h)$.

$= (x+h)^2 = (x+h) \cdot (x+h)$ — Expand $(x+h)^2$.

$= x^2 + hx + hx + h^2$

$= x^2 + 2hx + h^2$

$$\frac{[(x^2 + 2hx + h^2) - 2(x+h) + 5] - [x^2 - 2x + 5]}{h}$$

Substitute for $f(x+h)$ and $f(x)$ in the Difference Quotient.

$$\frac{x^2 + 2hx + h^2 - 2x - 2h + 5 - x^2 + 2x - 5}{h}$$

Distribute −2 and the minus sign.

$$\frac{2hx + h^2 - 2h}{h}$$

Combine like terms.

$$\frac{h(2x + h - 2)}{h}$$

Factor h out of the numerator.

$2x + h - 2$ Reduce.

As h approaches 0, the Difference Quotient approaches $2x - 2$. At the point (3, 8), the slope of the tangent line is

$$\begin{aligned} 2x - 2 &= 2(3) - 2 \\ &= 6 - 2 \\ &= 4 \end{aligned}$$

The equation of the tangent line is found using the point-slope formula.

$y - y_1 = m(x - x_1)$	Write the point-slope formula.
$y - 8 = 4(x - 3)$	Substitute $x_1 = 3$, $y_1 = 8$, and $m = 4$.
$y = 4x - 4$	Simplify.

The graph of the function and the tangent line follows.

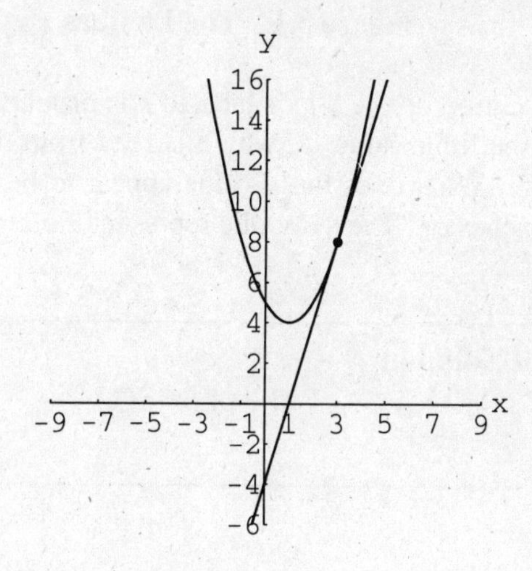

8.2 LIMITS

In the previous section, we examined the Difference Quotient as h approached 0. We will now formalize this idea. In order to do this we must begin with a discussion of limits. In this section, we will focus on:

(a) The basic concepts of limits
(b) Graphic interpretation of limits
(c) Algebraic techniques used in determining limits
(d) Finding limits at infinity
(e) Checking for continuity
(f) Limit definition of the derivative

The Concept of Limits

A graph can help clarify how limits are viewed in calculus. Consider the graph of a function, $f(x)$, in the first quadrant.

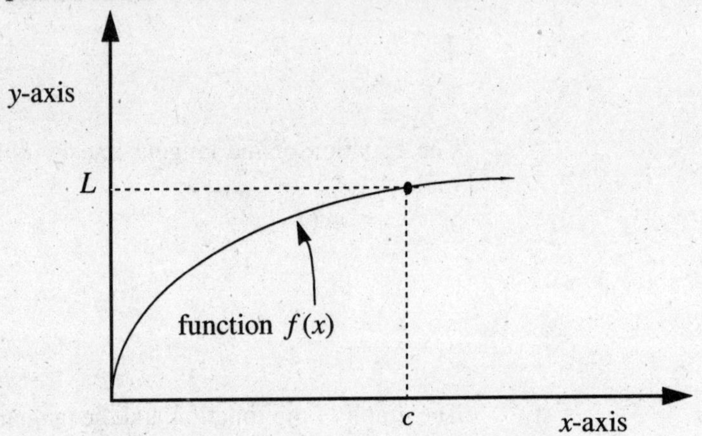

Figure 8.1 The Limit as x approaches c

The location on the x-axis labeled c is matched with the function to determine the limit value, L, which is read from the y-axis. Basically, we are asking, "What does the y value appear to be approaching as the x value approaches c?" The symbolic representation for a limit under these conditions is:

Limit Notation

$$\lim_{x \to c} f(x) = L$$

where:

lim is read "the limit"
$x \to c$ is read "as x approaches c"
$f(x)$ is any given function ($x^2 - 1$, for example)
L is the limit value

If numbers are substituted for c and L and a function substituted for $f(x)$, we could, for example, have

$$\lim_{x \to 2} (x^2 - 1) = 3$$

The example would read "the limit, as x approaches 2, of $x^2 - 1$ is 3." Note that when 2 is substituted into $x^2 - 1$, we have $2^2 - 1 = 4 - 1 = 3$ (the limit). In many cases, direct substitution of the c-value into a given function will give the limit, L.

Other aspects of the determination of limits can be shown by looking at the graph of

$$f(x) = \frac{x^3 - 3x^2 + x - 3}{x - 3}$$

Figure 8.2 shows this graph with an open circle at the (x, y) location of (3, 10).

This open circle indicates that this function does not exist at (3, 10), because a substitution of $x = 3$ in this function causes division by zero.

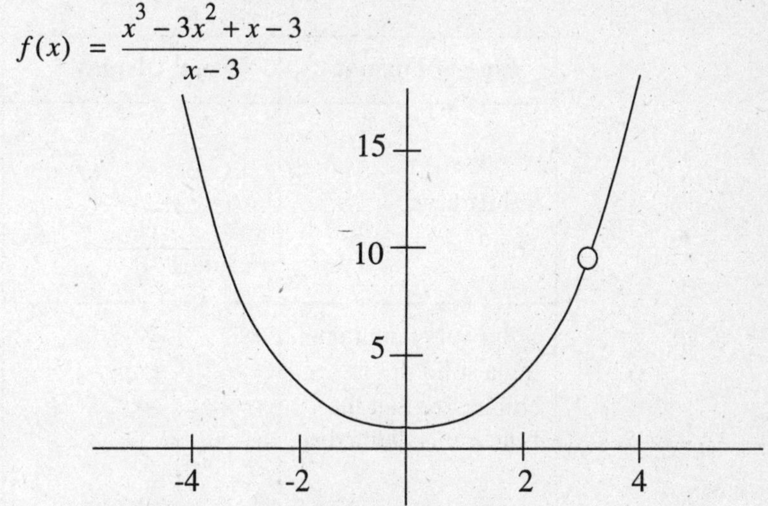

Figure 8.2 The Function Does Not Exist at (3, 10)

The graph shows that as x values get close to 3, $f(x)$ values get close to 10. We can state this in symbolic terms as

$$\lim_{x \to 3} \frac{x^3 - 3x^2 + x - 3}{x - 3} = 10$$

Even though we cannot actually let $x = 3$, we can choose values very near 3 to get an idea of what value the function approaches as x approaches 3. For example, let's substitute $x = 2.99$ and $x = 3.01$ in the function

$$f(x) = \frac{x^3 - 3x^2 + x - 3}{x - 3}$$

$$f(2.99) = \frac{(2.99)^3 - 3(2.99)^2 + (2.99) - 3}{2.99 - 3} = 9.99$$

$$f(3.01) = \frac{(3.01)^3 - 3(3.01)^2 + (3.01) - 3}{3.01 - 3} = 10.01$$

As you can see, the values are very close to 10.

Evaluating Limits Using Graphs

Many texts present the graph of a function and ask you to determine limit values at various points along the x-axis. Table 8.2 summarizes the common formats found in graphs and their meaning. Each format is examined for a function value and the limit value.

Type of Format	Visual Display	Meaning
Solid curve	c	Limit exists at c, Function exists at c.
Solid curve containing a solid dot to emphasize that the function is defined at c	c	Limit exists at c, Function exists at c.
Solid curve containing an open circle	c	Limit exists at c, Function is not defined at c.
Solid curve containing an open circle. Solid dot above or below open circle.	c	Limit exists and equals the y-value across from the open dot at c. Function exists and equals the y-value across from the solid dot.

Table 8.2 Graphic Representation of Limits

Type of Format	Visual Display	Meaning
Two solid curves terminating at different levels above c.	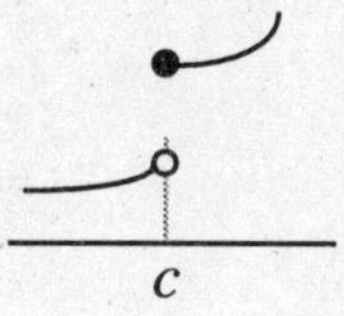 c	Limit does not exist at c. (We do not have the same y-value if c is approached from the left and right.) Function is defined and equals the y-value across from the solid dot.
Vertical asymptote occurs at c.	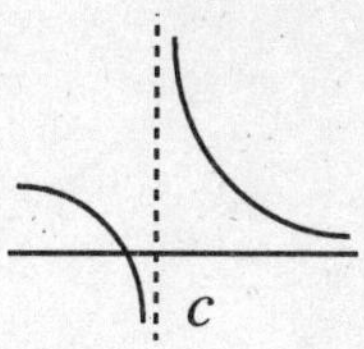 c	Function is not defined at c. Limit does not exist at c.

Table 8.2 (continued)

We must remember that the value of c on the x-axis can be approached by x from either the positive or the negative directions. Therefore, it is important to trace the function from *both* directions as we focus on the graph directly above point c on the x-axis.

One final point before we begin an exercise on determining limit values and function values. Students often have trouble with the difference between the limit of $f(x)$ as x approaches c and evaluating $f(x)$ at $x = c$ (to give $f(c)$).

Consider the following diagram with the open circle on a function "magnified."

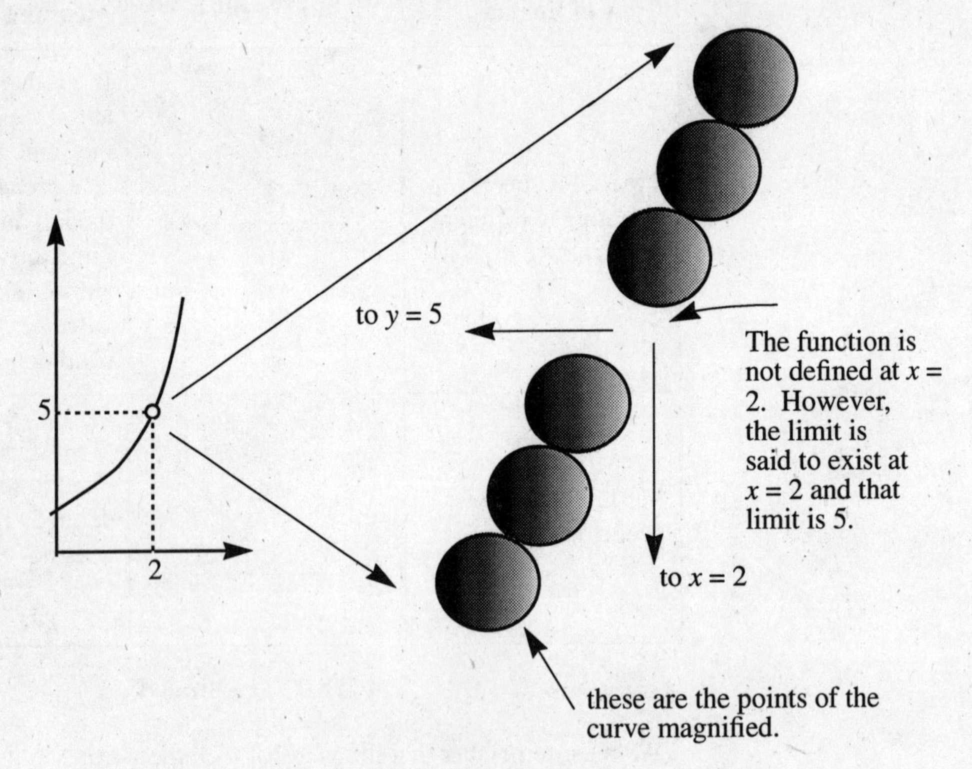

Figure 8.3 Magnification of a Limit

Study Figure 8.3. We know that the function is not defined at $x = 2$. However, we say that the limit of the function is 5 when x approaches 2.

EXAMPLE 8.4

Determine the values for the given functions in the graph that follows.

The function, $f(x)$:

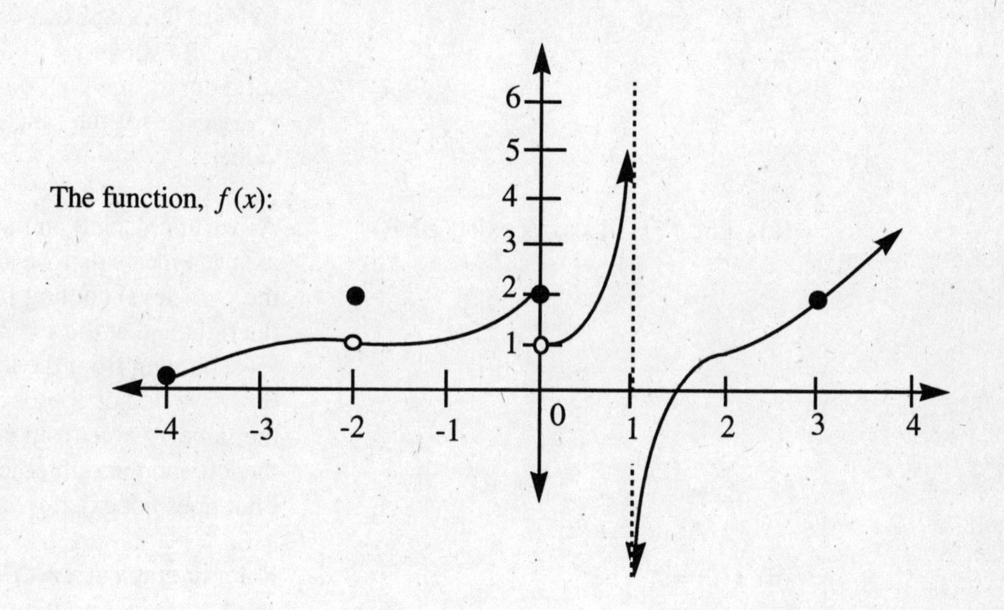

Find:

(a) $\lim_{x \to -2} f(x)$ (b) $f(-2)$

(c) $\lim_{x \to 0} f(x)$ (d) $f(0)$

(e) $\lim_{x \to 1} f(x)$ (f) $f(1)$

(g) $\lim_{x \to 2} f(x)$ (h) $f(2)$

(i) $\lim_{x \to 3} f(x)$ (j) $f(3)$

(k) $\lim_{x \to -4} f(x)$ (l) $f(-4)$

SOLUTION 8.4

(a) $\lim_{x \to -2} f(x) = 1$

Refer to the graph at $x = -2$. As we approach -2 on the x-axis from either direction, it appears as though we are approaching the y-value of 1.

(b) $f(-2)=2$

Refer to the graph at $x=-2$. Vertically above $x=-2$ is a solid dot even with 2 on the y-axis. This is the function value.

(c) $\lim_{x\to 0} f(x)$ does not exist (DNE)

As we approach 0 on the x-axis, it appears that we are at the $y=1$ level coming from the right and at the $y=2$ level coming from the left. Since we do not approach the same y-value from both the left and the right, the limit does not exist.

(d) $f(0)=2$

Refer to graph at $x=0$. The solid dot is at 2 on the y-axis.

(e) $\lim_{x\to 1} f(x)$ does not exist (DNE)

Refer to the graph at $x=1$. The limit does not exist since we do not get the same y-value as we approach from the left and the right.

(f) $f(1)$ does not exist (DNE)

Refer to the graph at $x=1$. A function is not defined at a vertical asymptote.

(g) $\lim_{x\to 2} f(x) = 1$

Refer to the graph at $x=2$. We are approaching the y-value of 1 when we approach $f(x)$ from either direction.

(h) $f(2)=1$

Refer to the graph at $x=2$. The function is defined at $y=1$ when $x=2$. No solid

	dot is necessary (although often a solid dot is used to emphasize that the function is defined at a given value).
(i) $\lim_{x \to 3} f(x) = 2$	Refer to the graph at $x = 3$. As we travel along the function approaching $x = 3$ from either direction, it appears we are at the level where $y = 2$.
(j) $f(3) = 2$	Refer to the graph at $x = 3$. The y-value of the function at $x = 3$ is $y = 2$.
(k) $\lim_{x \to -4} f(x)$ does not exist (DNE)	Refer to the graph at $x = -4$. We cannot approach $x = -4$ from the left at this location because the function does not exist to the left of $x = -4$.
(l) $f(-4) = 0.5$	Refer to the graph at $x = -4$. The function is at about the $y = 0.5$ level at $x = -4$.

Evaluating Limits Algebraically

While the visual inspection of a graph is one way to detemine limits, another process involves the use of algebraic techniques. It is usually best to try direct substitution first. If direct substitution produces division by 0, try one of the other algebraic techniques listed in Table 8.3. Also, recall from Example 8.4 that not all limits exist.

Technique	**Example**
Direct substitution	$\lim_{x \to 2} (3x - 5) = 3(2) - 5 = 1$
Removing a common factor, then trying substitution	$\dfrac{x^3 - 2x^2 + x}{x} = \dfrac{x(x^2 - 2x + 1)}{x} = x^2 - 2x + 1$

Table 8.3 Algebraic Techniques used in Determining Limits

Technique	Example
Factoring the difference of two squares	$\frac{x^2-9}{x-3} = \frac{(x+3)(x-3)}{(x-3)}$ $= x+3$
Rationalizing the numerator	$\frac{x-\sqrt{3}}{x^2-3} \cdot \frac{x+\sqrt{3}}{x+\sqrt{3}}$ $= \frac{(x^2-3)}{(x^2-3)(x+\sqrt{3})} = \frac{1}{x+\sqrt{3}}$
Factoring a trinomial	$\frac{x^2+7x+10}{x+2}$ $= \frac{(x+5)(x+2)}{(x+2)} = x+5$

Table 8.3 (continued)

EXAMPLE 8.5

Find the limit for the following functions.

(a) $\lim_{x \to 2} (3x^2 - 7)$

(b) $\lim_{x \to 0} \frac{x^2 - x}{\sqrt{x}}$

(c) $\lim_{x \to 5} \frac{x^2 - 25}{x - 5}$

(d) $\lim_{x \to 2} \frac{\sqrt{x} - \sqrt{2}}{x - 2}$

(e) $\lim_{x \to -3} \frac{2x^2 + 7x + 3}{x + 3}$

(f) $\lim_{x \to 10} 5$

(g) $\lim_{x \to 7} \frac{4}{7 - x}$

SOLUTION 8.5

(a) $\lim_{x \to 2} (3x^2 - 7)$ — Copy the function.

$3(2)^2 - 7 = 3(4) - 7 = 12 - 7 = 5$ — Use substitution.

$\lim_{x \to 2} (3x^2 - 7) = 5$ — The limit equals 5.

(b) $\lim_{x \to 0} \frac{x^2 - x}{\sqrt{x}}$ — Copy the function.

$\frac{0^2 - 0}{\sqrt{0}} = \frac{0}{0}$ undefined — Substitution does not yield a solution.

$\lim_{x \to 0} \frac{x^2 - x}{\sqrt{x}} = \lim_{x \to 0} \frac{x^2 - x}{x^{1/2}}$ — The limit cannot be evaluated when the function is in this form. Remove common factor.

$= \lim_{x \to 0} \frac{x^{1/2} x^{3/2} - x^{1/2} x^{1/2}}{x^{1/2}}$

$= \lim_{x \to 0} \frac{x^{1/2} (x^{3/2} - x^{1/2})}{x^{1/2}}$ — Remove $x^{1/2}$ and cancel.

$= \lim_{x \to 0} (x^{3/2} - x^{1/2}) =$

$= 0^{3/2} - 0^{1/2} = 0$ — The limit is 0.

(c) $\lim_{x \to 5} \frac{x^2 - 25}{x - 5}$ — Copy the function.

$\frac{5^2 - 25}{x - 5} = \frac{25 - 25}{5 - 5} = \frac{0}{0}$ undefined — The limit cannot be evaluated when the function is in this form.

$\lim_{x \to 5} \frac{x^2 - 25}{x - 5} = \lim_{x \to 5} \frac{(x - 5)(x + 5)}{(x - 5)}$ — Factor the numerator as a difference of two squares.

$\lim_{x \to 5} \frac{\cancel{(x - 5)}(x + 5)}{\cancel{(x - 5)}}$ — Cancel common factors.

$$\lim_{x \to 5} (x+5) = 5+5 = 10$$

The limit can now be evaluated by direct substitution.

(d) $$\lim_{x \to 2} \frac{\sqrt{x}-\sqrt{2}}{x-2}$$

Copy the function.

$$\frac{\sqrt{2}-\sqrt{2}}{2-2} = \frac{0}{0} \text{ undefined}$$

The limit cannot be evaluated when the function is in this form.

$$= \lim_{x \to 2} \frac{\sqrt{x}-\sqrt{2}}{x-2} \cdot \frac{\sqrt{x}+\sqrt{2}}{\sqrt{x}+\sqrt{2}}$$

Rationalize the numerator. The denominator must also be multiplied, but do not distribute in the denominator.

$$= \lim_{x \to 2} \frac{(x-2)}{(x-2)(\sqrt{x}+\sqrt{2})}$$

$$= \lim_{x \to 2} \frac{\cancel{(x-2)}}{\cancel{(x-2)}(\sqrt{x}+\sqrt{2})}$$

Cancel the common factors.

$$= \lim_{x \to 2} \frac{1}{\sqrt{x}+\sqrt{2}} = \frac{1}{\sqrt{2}+\sqrt{2}}$$

Evaluate the limit by direct substitution.

$$= \frac{1}{2\sqrt{2}}$$

The limit is now found.

(e) $$\lim_{x \to -3} \frac{2x^2+7x+3}{x+3}$$

Copy the function.

$$\frac{2(-3)^2+7(-3)+3}{-3+3} = \frac{18-21+3}{0}$$

Check to see if the limit exists at the given x-value.

$$= \frac{21-21}{0} = \frac{0}{0}$$

The limit cannot be evaluated when the function is in this form.

$$\lim_{x \to -3} \frac{2x^2+7x+3}{x+3}$$

$$= \lim_{x \to -3} \frac{(2x+1)(x+3)}{(x+3)}$$

Factor the numerator.

$$= \lim_{x \to -3} \frac{(2x+1)\,\cancel{(x+3)}}{\cancel{(x+3)}}$$ Cancel common factors.

$$= \lim_{x \to -3} (2x+1) = 2(-3)+1$$ The limit can now be found by direct substitution.

$$= -6+1 = -5$$

(f) $\lim_{x \to 10} 5$ Copy the function.

$\lim_{x \to 10} 5 = 5$ This function is a constant so the limit for this function will always be 5, regardless of what value x approaches.

(g) $\lim_{x \to 7} \frac{4}{7-x}$ Copy the function.

$\frac{4}{7-7} = \frac{4}{0}$ undefined The function has no limit at this x-value.

There is no algebraic technique available to convert this function into a form to obtain a limit. Many students find it difficult to determine when algebra works and when it does not. The best solution is to practice with many problems.

Limits at Infinity

To describe the situation where the x-values increase or decrease without bound, we use the following notation:

$$\lim_{x \to \infty} f(x) \text{ and } \lim_{x \to -\infty} f(x)$$

If the y-value approaches a number L in this case, we say that the limit as x approaches infinity (or negative infinity) exists and equals L.

There is a general rule involving a reciprocal power of x that helps us evaluate limits at infinity. The formula is:

$\lim_{x \to \infty} \frac{1}{x^p} = 0$ where p is any real number greater than zero.

This formula allows us to evaluate an expression like:

$$\lim_{x \to \infty} \frac{2x^2 - x}{3x^2 + 1}$$

by dividing each of the terms by the highest power of x (x^2 in this case.)

The function now becomes:

$$\lim_{x \to \infty} \frac{\frac{2x^2}{x^2} - \frac{x}{x^2}}{\frac{3x^2}{x^2} + \frac{1}{x^2}} = \lim_{x \to \infty} \frac{2 - \frac{1}{x}}{3 + \frac{1}{x^2}}$$

The terms $1/x$ and $1/x^2$ are essentially zero when x is very large (divide 1 by successively larger numbers and see how the quotient becomes very small). The limit is now found.

$$\lim_{x \to \infty} \frac{2 - \frac{1}{x}}{3 + \frac{1}{x^2}} = \frac{2 - 0}{3 + 0} = \frac{2}{3}$$

The process of dividing by the largest power of x may be time-consuming. This process can be shortened by observing each of the three cases that occur with this type of problem.

Case	Highest Power	Limit	Example
I	In the numerator	$\pm\infty$	$\lim_{x \to \infty} \frac{x^3 - 1}{x^2} = \infty$
II	Equal in both numerator and denominator	Ratio of coefficients of highest power	$\lim_{x \to \infty} \frac{3x^2 - 1}{7x^2 - x} = \frac{3}{7}$

Table 8.4 Limits at Infinity for Rational Functions

Case	Highest Power	Limit	Example
III	In the denominator	0	$\lim_{x \to \infty} \frac{5-x}{x^3+2} = 0$

Table 8.4 (continued)

EXAMPLE 8.6

Find the following limits.

(a) $\lim_{x \to \infty} \frac{x^3+2x-1}{x^2+8x}$

(b) $\lim_{x \to \infty} \frac{x+1}{x^3+2}$

(c) $\lim_{x \to \infty} \frac{8x^2+x}{11x^2+2}$

(d) $\lim_{x \to -\infty} \frac{x^3+x}{x^2-2x}$

SOLUTION 8.6

(a) $\lim_{x \to \infty} \frac{x^3+2x-1}{x^2+8x}$ Copy the function.

$\lim_{x \to \infty} \frac{x^3+2x-1}{x^2+8x} = \infty$ Since the highest power is in the numerator, the limit is ∞.

(b) $\lim_{x \to \infty} \frac{x+1}{x^3+2}$ Copy the function.

$\lim_{x \to \infty} \frac{x+1}{x^3+2} = 0$ The highest power is in the denominator. The limit is zero.

(c) $\lim_{x \to \infty} \frac{8x^2+x}{11x^2+2}$ Copy the function.

$$\lim_{x \to \infty} \frac{8x^2 + x}{11x^2 + 2} = \frac{8}{11}$$

The largest exponents are the same in the numerator and the denominator. The limit is the ratio of the coefficients.

(d) $$\lim_{x \to -\infty} \frac{x^3 + x}{x^2 - 2x}$$

Copy the function.

$$\lim_{x \to -\infty} \frac{x^3 + x}{x^2 - 2x}$$

The largest exponent is in the numerator.

$$\lim_{x \to -\infty} \frac{x^3 + x}{x^2 - 2x} = -\infty$$

The value of x^3 is *negative* and the value of x^2 is *positive* as x approaches $-\infty$. The quotient is *negative*.

Continuity

The concept of a continuous function is that it follows a smooth uninterrupted path. The function cannot have any breaks, vertical asymptotes, or "holes." There is a formal definition for continuity. A function is continuous at c if:

(I)	$\lim_{x \to c} f(x) = L$	(the limit exists)
(II)	$f(c)$ exists	(the function value at c exists)
(III)	$f(c) = L$	(the function value at c equals the limit)

EXAMPLE 8.7

The graph of a function, $f(x)$, follows. Determine whether the function is continuous at each value.

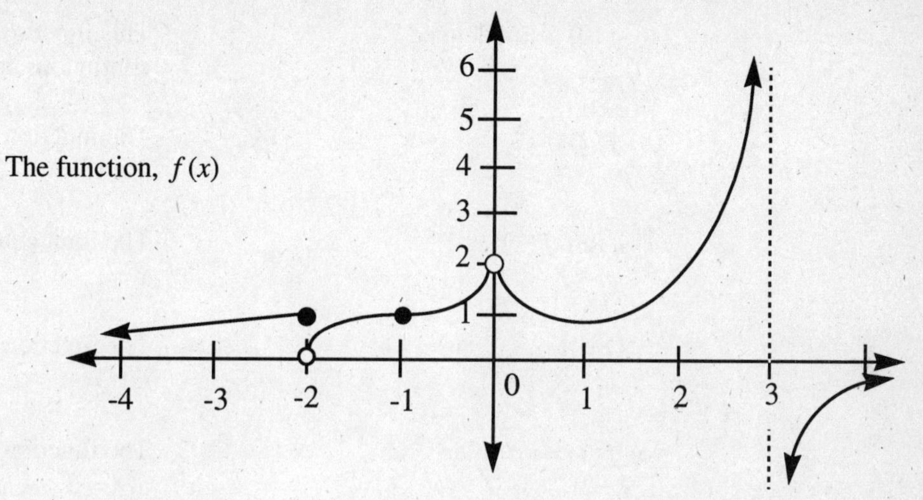

(a) at $x = -2$

(b) at $x = -1$

(c) at $x = 0$

(d) at $x = 1$

(e) at $x = 3$

SOLUTION 8.7

(a) $f(-2) = 1$ — The function at -2 is 1.

$\lim_{x \to -2} f(x)$ does not exist — The limit does not exist.

$\lim_{x \to -2} f(x) \neq 1$ — This function is *not* continuous at $x = -2$.

(b) $f(-1) = 1$ — The function equals 1 at $x = -1$.

$\lim_{x \to -1} f(x) = 1$ — The limit is 1 at $x = -1$. Therefore, the function is continuous at $x = -1$.

(c) $f(0) =$ does not exist — There is *no* value for this function at $x = 0$.

$\lim_{x \to 0} f(x) = 2$	The limit at $x = 0$ is 2.
$f(0)$ is not defined	This function is *not* continuous at $x = 0$.
(d) $f(1) = 1$	The function equals 1 at $x = 1$.
$\lim_{x \to 1} f(x) = 1$	The limit equals 1 at $x = 1$.
$1 = 1$	The function is continuous at $x = 1$.
(e) $f(3) =$ no value	The function is not defined at $x = 3$.
$\lim_{x \to 3} f(x)$ does not exist	The limit does not exist at $x = 3$.

The function value at 3 and the limit as x approaches 3 do not exist. The function is not continuous at $x = 3$.

Limit Definition of the Derivative

We can now return to the discussion from section 8.1 where we allowed h to approach 0 in the Difference Quotient. We now know that we can write that as

$$\lim_{h \to 0} \frac{f(x+h) - f(x)}{h}$$

In fact, this is precisely the limit definition of the derivative of f at x, and is written as $f'(x)$.

Derivative of *f*

$$f'(x) = \lim_{h \to 0} \frac{f(x+h) - f(x)}{h}$$

EXAMPLE 8.8

Find the derivative of $f(x) = 6 - 2x^2$.

SOLUTION 8.8

Find the Difference Quotient:

$$\frac{f(x+h) - f(x)}{h} = \frac{(6 - 2(x+h)^2) - (6 - 2x^2)}{h}$$

Replace x with $(x + h)$ and subtract $f(x)$.

$$= \frac{6 - 2x^2 - 4xh - 2h^2 - 6 + 2x^2}{h}$$

Multiply and distribute.

$$= \frac{-4xh - 2h^2}{h}$$

Add similar terms.

Now evaluate the limit as h approaches 0 by first factoring (since direct substitution will yield division by 0):

$$\lim_{h \to 0} \frac{-4xh - 2h^2}{h} = \lim_{h \to 0} \frac{h(-4x - 2h)}{h}$$

Factor out h.

$$= \lim_{h \to 0} (-4x - 2h)$$

Reduce.

$$= -4x$$

Substitute 0 for h.

8.3 RULES OF DIFFERENTIATION

Notation

In section 8.2 we introduced $f'(x)$ as the notation for derivative. The following notation may also be used to signify a derivative of $f(x)$:

Function Notation	**Derivative Notation**	**Name**
$f(x)$	$f'(x)$	function notation
y	y'	y-prime
y	$\frac{dy}{dx}$	Leibniz notation
an expression	D_x	operator notation

Table 8.5 Derivative Symbols

Derivative Shortcuts

Using the limit definition of a derivative is time-consuming. Fortunately, there are several shortcuts (proofs are available in any standard calculus textbook). The rules that follow should be memorized.

Function	Derivative	Derivative Rules
$f(x) = 5$	$f'(x) = 0$	The derivative of a constant is 0.
$f(x) = x^4$	$f'(x) = 4x^3$	The derivative of x^n is nx^{n-1} (called the Power Rule.
$f(x) = 5x^4$	$f'(x) = 5(x^4)'$ $= 5(4x^3)$ $= 20x^3$	The derivative of a constant times a function is the constant times the derivative of the function.
$f(x) = x^2 + x^5$	$f'(x) = 2x + 5x^4$	The derivative of a sum or difference of functions is the sum or difference of the derivatives.

Table 8.6 Derivative Rules

Note that rules for finding derivatives of products and quotients will be introduced in the next chapter.

EXAMPLE 8.9

Find the derivative of each function.

(a) $f(x) = 5$

(b) $f(x) = 3x$

(c) $f(x) = 7x^2 - 3x + 5$

(d) $f(x) = \sqrt{x}$

(e) $f(x) = 5\sqrt{x} + \dfrac{7}{\sqrt[3]{x}}$

(f) $f(x) = \sqrt[3]{x^5} + 8x^7$

(g) $f(x) = 30x^6$

(h) $f(x) = 3x^{\sqrt{2}}$

(i) $f(x) = \pi$

(j) $f(x) = ax^b$ where a and b are real numbers.

SOLUTION 8.9

(a) $f(x) = 5$ — Copy the function.

$f'(x) = 0$ — The derivative of a *constant* is *zero*.

(b) $f(x) = 3x$ — Copy the function.

$f(x) = 3x^1$ — Express the understood exponent of 1.

$f'(x) = 3 \cdot 1x^{1-1}$ — Use the Power Rule.

$= 3x^0 = 3 \cdot 1 = 3$ — Simplify.

(c) $f(x) = 7x^2 - 3x + 5$ — Copy the function.

$= 7x^2 - 3x^1 + 5$ — Express the exponents for clarity.

$f'(x) = 2 \cdot 7x^{2-1} - 1 \cdot 3x^{1-1} + 0$ — Use the Power Rule and the fact that the derivative of a constant is 0.

$f'(x) = 14x - 3x^0 + 0$

$f'(x) = 14x - 3$ — Simplify.

(d) $f(x) = \sqrt{x}$	Copy the function.
$f(x) = x^{1/2}$	Rewrite the radical in fractional exponent form.
$f'(x) = \frac{1}{2}x^{1/2 - 1}$	Use the Power Rule.
$f'(x) = \frac{1}{2}x^{1/2 - 2/2}$	
$f'(x) = \frac{1}{2}x^{-1/2}$	
$f'(x) = \frac{1}{2} \cdot \frac{1}{x^{1/2}} = \frac{1}{2} \cdot \frac{1}{\sqrt{x}}$	Rewrite the derivative in radical form.
$= \frac{1}{2\sqrt{x}}$	
(e) $f(x) = 5\sqrt{x} - \frac{7}{\sqrt[3]{x}}$	Copy the function.
$f(x) = 5x^{1/2} - \frac{7}{x^{1/3}}$	Rewrite the radicals as fractional exponents.
$f(x) = 5x^{1/2} - 7x^{-1/3}$	Convert the last term from a fraction to a term with a negative exponent.
$f'(x) = 5\left(\frac{1}{2}\right)x^{\frac{1}{2} - 1} - 7\left(-\frac{1}{3}\right)x^{-\frac{1}{3} - 1}$	Use the Power Rule.
$f'(x) = \frac{5}{2}x^{-\frac{1}{2}} + \frac{7}{3}x^{-\frac{4}{3}}$	Simplify the terms.
$f'(x) = \frac{5}{2}\frac{1}{x^{1/2}} + \frac{7}{3}\frac{1}{x^{4/3}}$	Rewrite negative exponents as reciprocals.
$f'(x) = \frac{5}{2\sqrt{x}} + \frac{7}{3\sqrt[3]{x^4}}$	Convert fractional exponents into radical form.
(f) $f(x) = \sqrt[3]{x^5} + 8x^7$	Copy the function.
$f(x) = x^{5/3} + 8x^7$	Rewrite the radical in fractional exponent form.

$f'(x) = \frac{5}{3}x^{\frac{5}{3}-1} + 8(7)x^{7-1}$ Use the Power Rule.

$f'(x) = \frac{5}{3}x^{2/3} + 56x^6$ Simplify the terms.

$f'(x) = \frac{5}{3}\sqrt[3]{x^2} + 56x^6$ Convert the fractional exponent to radical form.

(g) $f(x) = 30x^6$ Copy the function.

$f'(x) = 6(30)x^{6-1}$ Use the Power Rule.

$f'(x) = 180x^5$ Simplify the function.

(h) $f(x) = 3x^{\sqrt{2}}$ Copy the function.

$f'(x) = 3(\sqrt{2})(x^{\sqrt{2}-1})$ Use the Power Rule.

$f'(x) = 3\sqrt{2}x^{\sqrt{2}-1}$ Simplify the function as much as possible.

(i) $f(x) = \pi$ Copy the function.

$f'(x) = 0$
because π is a constant.

(j) $f(x) = ax^b$ Copy the function.

$f'(x) = (a)(b)x^{b-1}$ Use the Power Rule. (Remember a and b are numbers.)

$f'(x) = abx^{b-1}$

EXAMPLE 8.10

Find an equation of the tangent line to $f(x) = x^3 + 2x + 1$ at (1, 4).

SOLUTION 8.10

Since the slope of the tangent line at $x = 1$ equals the derivative at $x = 1$, find $f'(x)$ at $x = 1$.

$f'(x) = 3x^2 + 2$ Use the Power Rule.

We need the slope at $x = 1$.

$f'(1) = 3(1)^2 + 2 = 5$

Now use the point-slope form of a line:

$y - y_1 = m(x - x_1)$

$y - 4 = 5(x - 1)$ Substitute $(x_1, y_1) = (1, 4)$ and $m = 5$.

$y - 4 = 5x - 5$ Simplify.
$y = 5x - 1$

8.4 THE CHAIN RULE

As functions get more complex in nature, it may be useful to use a more versatile rule. The General Power Rule (or Chain Rule) is useful for composite functions.

> **The General Power Rule**
>
> Given: $f(x) = [g(x)]^n$
>
> Then: $f'(x) = n[g(x)]^{n-1}(g'(x))$

Note that the function $g(x)$ is not changed in the first part of the power rule. However, after subtracting 1 from the power, multiply by the derivative of g.

For example, the following function has a binomial $(x^2 + 3x)$ raised to the 4th power:

$$f(x) = (x^2 + 3x)^4$$

Finding the derivative of this function begins by treating the function as if it were a variable to the 4th power. After the exponent has been moved according to the Power Rule, then the binomial inside the parentheses is differentiated. Table 8.7 shows the thought process involved in the differentiation of the entire function.

$f(x) = (x^2 + 3x)^4$	Begin with the Power Rule. Do *not* focus on contents of parentheses.
$f'(x) = \mathbf{4}(\quad)^{\mathbf{3}}$	
$f'(x) = \mathbf{4}(x^2 + 3x)^{\mathbf{3}}$	*Now* turn your attention to the binomial.
$f'(x) = \mathbf{4}(x^2 + 3x)^{\mathbf{3}} \cdot (\mathbf{2x + 3})$ — derivative of $x^2 + 3x$	Multiply by the derivative of the binomial.
$f'(x) = 4(2x + 3)(x^2 + 3x)^3$	Rearranging the factors completes the process of finding the derivative.

Table 8.7 The General Power Rule

EXAMPLE 8.11

Find the derivative of the following functions. Use the General Power Rule.

(a) $f(x) = (3x^2 + 7x + 1)^5$

(b) $f(x) = 4(5 - x)^6$

(c) $f(x) = \sqrt[3]{x + 1}$

(d) $f(x) = \dfrac{5}{x^2 + 2x + 4}$

(e) $f(x) = \dfrac{3}{\sqrt{x^2 + 2x}}$

SOLUTION 8.11

(a) $f(x) = (3x^2 + 7x + 1)^5$ Copy the function.

$f(x) = (3x^2 + 7x + 1)^{\mathbf{5}} = [g(x)]^{\mathbf{5}}$	Note that this function is an expression to the 5th power.
$f'(x) = \mathbf{5}\,(3x^2 + 7x + 1)^{\mathbf{4}}\,(\mathbf{6x + 7})$ $g(x) \qquad g'(x)$	Use the General Power Rule. The bold letters and numbers are changes through differentiation.
$f'(x) = 5\,(6x + 7)\,(3x^2 + 7x + 1)^4$	Move the binomial for clarity.
(b) $f(x) = 4\,(5 - x)^6$	Copy the function.
$f(x) = 4\,(5 - x)^{\mathbf{6}} = 4\,[g(x)]^{\mathbf{6}}$	Note that this function is a constant times an expression to the 6th power.
$f'(x) = 4 \cdot \mathbf{6}\,(5 - x)^{\mathbf{5}}\,(\mathbf{-1})$ $g(x) \qquad g'(x)$	Use the General Power Rule. The bold typeface shows changes.
$f'(x) = -24\,(5 - x)^5$	Simplify.
(c) $f(x) = \sqrt[3]{x + 1}$	Copy the function.
$f(x) = (x + 1)^{1/3}$	Rewrite the radical as a fractional exponent.
$f(x) = (x + 1)^{\mathbf{1/3}} = [g(x)]^{\mathbf{1/3}}$	Note that this function is an expression to the $1/3$ power.
$f'(x) = \frac{1}{3}\,(x + 1)^{\frac{1}{3} - 1}\,(1)$ $= \mathbf{\frac{1}{3}}\,(x + 1)^{\mathbf{-2/3}}\,(\mathbf{1})$ $g(x) \qquad g'(x)$	Use the General Power Rule. The bold typeface shows changes.
$f'(x) = \frac{1}{3}\,(x + 1)^{-\frac{2}{3}}$	Simplify the function.
$f'(x) = \frac{1}{3}\,\frac{1}{\sqrt[3]{(x + 1)^2}} = \frac{1}{3\sqrt[3]{(x + 1)^2}}$	Rewrite as a radical

expression.

(d) $f(x) = \dfrac{5}{x^2 + 2x + 4}$ — Copy the function.

$f(x) = 5(x^2 + 2x + 4)^{-1}$ — Rewrite the fraction as an expression with a negative exponent.

$f(x) = 5(x^2 + 2x + 4)^{\mathbf{-1}} = 5[g(x)]^{\mathbf{-1}}$

Note that the function is a constant times an expression to the –1 power.

$f'(x) = (5)(\mathbf{-1})(x^2 + 2x + 4)^{-1\,\mathbf{-1}}(\mathbf{2x + 2})$

Use the general power rule. The bold typeface shows the changes.

$f'(x) = -5(2x + 2)(x^2 + 2x + 4)^{-2}$ — Simplify the function.

$f'(x) = \dfrac{-5(2x + 2)}{(x^2 + 2x + 4)^2}$ — Rewrite the expression as a rational function.

$f'(x) = \dfrac{-10(x + 1)}{(x^2 + 2x + 4)^2}$ — Factor the numerator.

(e) $f(x) = \dfrac{3}{\sqrt{x^2 + 2x}}$ — Copy the function.

$f(x) = \dfrac{3}{(x^2 + 2x)^{1/2}}$ — Rewrite the radical expression as an expression with a rational exponent.

$f(x) = 3(x^2 + 2x)^{-\frac{1}{2}}$ — Rewrite the fraction using a negative exponent.

$f(x) = 3[g(x)]^{\mathbf{-\frac{1}{2}}}$ — Note the function is a constant times an expression to the $-1/2$ power.

$$f'(x) = 3\left(-\frac{1}{2}\right)(x^2 + 2x)^{\mathbf{-\frac{1}{2}-1}}\ (\mathbf{2x+2})$$

Use the General Power Rule. The bold typeface shows the changes.

$$f'(x) = -\frac{3}{2}(x^2 + 2x)^{-\frac{3}{2}}(2x + 2)$$

$$f'(x) = -\frac{3}{2}(2x + 2)(x^2 + 2x)^{-\frac{3}{2}}$$

Simplify the expression.

$$f'(x) = (-3x - 3)(x^2 + 2x)^{-\frac{3}{2}}$$

$$f'(x) = \frac{-3x - 3}{\sqrt{(x^2 + 2x)^3}}$$

Rewrite the negative exponent as a radical expression.

8.5 DERIVATIVES OF LOGARITHMIC AND EXPONENTIAL FUNCTIONS

Both the natural logarithm (ln) and the natural number (e) have business, engineering, and science applications. They both also have unique derivatives. The basic derivatives for the functions e^x and $\ln x$ are listed in the rules below.

The Derivative of e^x

If $f(x) = e^x$ then $f'(x) = e^x$

The derivative of e^x, then, is simply e^x.

The Derivative of $\ln x$

If $f(x) = \ln x$ then $f'(x) = \frac{1}{x}$

These rules can be extended to situations where the exponent of e^x or the argument of $\ln x$ are not the single variable, x, but a function of x. We can then find the derivatives for $f(x) = e^{[g(x)]}$ and $f(x) = \ln[g(x)]$, where $g(x)$ is any function of x. The rules are given below.

Derivative for $e^{[g(x)]}$

If $f(x) = e^{[g(x)]}$ then $f'(x) = e^{[g(x)]} \cdot [g'(x)]$
where $g(x)$ is a function of x

The above rule can be loosely paraphrased by stating that the derivative of e^{function} is $(e^{\text{function}}) \cdot (\text{derivative of the function})$.

The rule for the derivative of the natural logarithm of a function follows a similar pattern.

Derivative of $\ln[g(x)]$

If $f(x) = \ln[g(x)]$ then $f'(x) = \dfrac{1}{g(x)} \cdot [g'(x)]$

Thus, the natural logarithm of a function has a derivative that is the reciprocal of the function times the derivative of the function.

EXAMPLE 8.12

Find the derivative of the following functions.

(a) $f(x) = 3e^{x}$

(b) $f(x) = 5\ln x$

(c) $f(x) = 2e^{x^2 - 1}$

(d) $f(x) = \ln(x^3 - 2x + 2)$

(e) $f(x) = e^{-x} + e^{x}$

(f) $f(x) = 3\ln x^2 + 5e^{x^2 - 2}$

SOLUTION 8.12

(a) $f(x) = 3e^x$ — Copy the function.

$f'(x) = 3e^x(\mathbf{1})$

$= 3e^x$

The numerical coefficient of e^x is not affected in the derivative process if the exponent is x. The derivative of x is 1 and $1 \cdot 3 = 3$.

(b) $f(x) = 5\ln x$ — Copy the function.

$f'(x) = 5\left(\frac{1}{x}\right)$

$f'(x) = \frac{5}{x}$

The derivative of the natural logarithm, $\left(\frac{1}{x}\right)$, is multiplied by 5.

(c) $f(x) = 2e^{x^2 - 1}$ — Copy the function.

$g(x) = x^2 - 1;\ g'(x) = 2x$ — Here, $g(x) = x^2 - 1$. Find the derivative.

$f'(x) = 2e^{x^2 - 1}(\mathbf{2x})$ — Multiply the original function by the derivative of $g(x)$ (bold type).

$f'(x) = 4xe^{x^2 - 1}$ — Simplify.

(d) $f(x) = \ln(x^3 - 2x + 2)$ — Copy the function.

$g(x) = x^3 - 2x + 2$ — Find the derivative of $g(x)$.

$g'(x) = 3x^2 - 2$

$f'(x) = \frac{\mathbf{1}}{\mathbf{x^3 - 2x + 2}} \cdot (\mathbf{3x^2 - 2})$ — Multiply the reciprocal of $g(x)$ by the derivative of $g(x)$.

$f'(x) = \frac{3x^2 - 2}{x^3 - 2x + 2}$ — Simplify.

(e) $f(x) = e^{-x} + e^{x}$	Copy the function.
$f'(x) = e^{-x}(\mathbf{-1}) + e^{x}$	Each term follows the rules for the derivative of $e^{g(x)}$.
$f'(x) = -e^{-x} + e^{x}$	Simplify.
(f) $f(x) = 3\ln x^2 + 5e^{x^2 - 2}$	Copy the function.
$f(x) = 3(2)\ln x + 5e^{x^2 - 2}$ $f(x) = 6\ln x + 5e^{x^2 - 2}$	Use the rules of logarithms to rewrite the first term.
$g(x) = x^2 - 2;\ g'(x) = 2x$	The first term is a constant times ln x, and the rule for the derivative of ln x applies. The second term has the function $(x^2 - 2)$ as the exponent of e. Let $g(x) = (x^2 - 2)$. Then $g'(x) = 2x$.
$f'(x) = 6\left[\mathbf{\frac{1}{x}}\right] + 5e^{x^2 - 2}[\mathbf{2x}]$	The bold type shows changes occurring through differentiation.
$f'(x) = \frac{6}{x} + 10xe^{x^2 - 2}$	Simplify.

Practice Exercises

1. For the following functions use the formula

$$\lim_{h \to 0} \frac{f(x+h) - f(x)}{h}$$

to find the slope of the tangent.

(a) $f(x) = 7x + 2$

(b) $f(x) = 3x^2 - x$

2. Find the limit for the following functions.

(a) $\lim_{x \to -1} (x^2 + x)$

(b) $\lim_{x \to 0} \frac{x^3 - x}{x}$

(c) $\lim_{x \to 0} \frac{\sqrt{x+2} - \sqrt{2}}{x}$

(d) $\lim_{x \to -3} \frac{x^2 + 5x + 6}{x + 3}$

(e) $\lim_{x \to 4} 3$

3. Find the following limits at infinity.

(a) $\lim_{x \to \infty} \frac{x^2 - 6x + 2}{x - 5}$

(b) $\lim_{x \to \infty} \frac{3x - 6}{7x + 4}$

(c) $\lim_{x \to \infty} \frac{x + 6}{x^2 - 7}$

4. Find the derivative of the following functions.

(a) $f(x) = 3x - 5$

(b) $f(x) = 6x^2 - 5x + 72$

(c) $f(x) = \sqrt{x^3}$

(d) $f(x) = 5\sqrt[3]{x} - 2x^{-2}$

(e) $f(x) = x^2$

(f) $f(x) = 1 + \sqrt{2}$

5. Use the General Power Rule to find the derivative of the following functions.

(a) $f(x) = (2x - 1)^6$

(b) $f(x) = 3(x^2 + 7x)^4$

(c) $f(x) = \frac{1}{(x-1)^2}$

(d) $f(x) = \frac{8}{\sqrt[3]{(x+1)^2}}$

6. Find the derivative of the following functions.

(a) $f(x) = \frac{1}{2}e^x$

(b) $f(x) = 2x + 4\ln x$

(c) $f(x) = e^{x^2 - 5}$

(d) $f(x) = \ln(x^3 - 2x^2 - 6)$

(e) $f(x) = \ln\left(\frac{e^x}{x}\right)$

(f) $f(x) = e^{6x + \ln x^2}$

Answers

1. (a) 7

   (b) $6x - 1$

2. (a) 0

   (b) -1

   (c) $\frac{1}{2\sqrt{2}} = \frac{\sqrt{2}}{4}$

   (d) -1

   (e) 3

3. (a) ∞

   (b) $3/7$

   (c) 0

4. (a) 3

   (b) $12x - 5$

   (c) $\frac{3}{2}x^{1/2}$

   (d) $\frac{5}{3}x^{-2/3} + 4x^{-3}$

   (e) $2x$

   (f) 0

5. (a) $12(2x-1)^5$

   (b) $12(2x+7)(x^2+7x)^3$

   (c) $\frac{-2}{(x-1)^3}$

   (d) $\frac{-16}{3\sqrt[3]{(x+1)^5}}$

6. (a) $\frac{1}{2}e^x$

   (b) $2 + \frac{4}{x}$

   (c) $2xe^{x^2-5}$

   (d) $\frac{3x^2 - 4x}{x^3 - 2x^2 - 6}$

   (e) $\frac{x-1}{x}$

   (f) $e^{6x + \ln x^2}\left(6 + \frac{2}{x}\right)$

9

More on Derivatives

In this chapter we will continue our discussion of derivatives by extending our rules to cover products and quotients of functions. We will also discuss higher-order derivatives and implicit differentiation.

9.1 THE PRODUCT AND QUOTIENT RULES

The derivative of a product of two functions is, unfortunately, generally *not* the product of their respective derivatives. Similarly, the derivative of a quotient is *not* the quotient of the derivatives. The rules for products and quotients are discussed in this section.

The Product Rule

The Product Rule

If $h(x) = f(x) \cdot g(x)$ then $h'(x) = f'(x) g(x) + f(x) g'(x)$
where
$f(x)$ = first factor and $g(x)$ = second factor.

The Product Rule can be stated in a less symbolic form. If the original function, $f(x)$, can be described as a product of two factors, then the derivative of the function, $f'(x)$, is

(first factor derivative)(second factor) + (first factor)(second factor derivative)

The function, $f(x) = (x^2 - 1)(3x^3 + 2x)$, can be differentiated using the product rule. It is often helpful to list each factor with its respective derivative immediately below it.

	First	Second
Factor:	$x^2 - 1$ $F(x)$	$3x^3 + 2x$ $S(x)$
Derivative of Factor:	$2x$ $F'(x)$	$9x^2 + 2$ $S'(x)$

Pair each derivative with the other factor to get the complete derivative of $f(x)$.

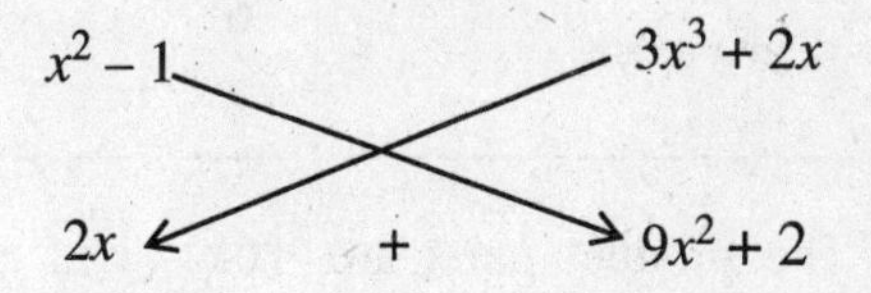

Therefore: $f'(x) = \underset{F'(x)}{[2x]}\ \underset{S(x)}{(3x^3 + 2x)} + \underset{F(x)}{(x^2 - 1)}\ \underset{S'(x)}{[9x^2 + 2]}$

The Quotient Rule

The Quotient Rule

$$\text{For } f(x) = \frac{t(x)}{b(x)}$$

$$f'(x) = \frac{t'(x)\,b(x) - t(x)\,b'(x)}{(b(x))^2}$$

where $t(x)$ is the numerator (top) function and $b(x)$ is the denominator (bottom) function.

The less symbolic statement for the quotient rule of $f(x)$ is

$$f'(x) = \frac{(\text{derivative of top})\ (\text{bottom}) - (\text{top})\ (\text{derivative of bottom})}{(\text{bottom})^{\text{squared}}}$$

The function, $f(x) = \dfrac{3x^2 + 7}{x^3 + 5x}$ can be differentiated using the quotient rule. As with the Product Rule, the numerator and the denominator can be listed with their respective derivatives. Note: Always list the numerator first.

	Numerator	**Denominator**
Function	$3x^2 + 7$ $t(x)$	$x^3 + 5x$ $b(x)$
Derivative	$6x$ $t'(x)$	$3x^2 + 5$ $b'(x)$

The product pairs are $(6x)\ (x^3 + 5x)$ and $(3x^2 + 7)\ (3x^2 + 5)$. The numerator of the **derivative** is $(6x)(x^3 + 5x) - (3x^2 + 7)(3x^2 + 5)$. It is formed by subtracting the product pairs in the *proper* order.

The derivative, which is itself a function, is

$$f'(x) = \frac{(6x)\ (x^3 + 5x) - (3x^2 + 7)\ (3x^2 + 5)}{(x^3 + 5x)^2}$$

After simplifying the numerator, the derivative becomes:

$$f'(x) = \frac{-(3x^4 + 6x^2 + 35)}{(x^3 + 5x)^2}$$

One of the most frustrating aspects of the Product Rule and the Quotient Rule is not the actual differentiation, but the factoring and simplifying needed after the differentiation is finished. In the next example, the simplification process will be given in detail.

EXAMPLE 9.1

Find the derivatives of the following functions. Use the General Power Rule where necessary.

(a) $f(x) = (x^2 + 1)(3x - 1)$

(b) $f(x) = \dfrac{(2x-3)}{(5x+1)}$

(c) $f(x) = (3x+1)^5 (x^2 - x)^4$

(d) $f(x) = \dfrac{(2x-5)^4}{(4x-2)^3}$

(e) $f(x) = \dfrac{e^x}{\ln x}$

SOLUTION 9.1

(a) $f(x) = (x^2 + 1)(3x - 1)$ — Copy the function.

Factor	$(x^2 + 1)$	$(3x - 1)$
Derivative	$(2x)$	(3)

List each factor and its derivative.

$f'(x) = (\mathbf{2x})(3x - 1) + (x^2 + 1)(\mathbf{3})$

Multiply each factor by the derivative of the other factor. Each derivative is in boldface type.

$f'(x) = 6x^2 - 2x + 3x^2 + 3$ — Multiply.

$f'(x) = 9x^2 - 2x + 3$ — Combine like terms.

(b) $f(x) = \dfrac{(2x-3)}{(5x+1)}$ — Copy the function.

	numerator	denominator
Factor	$(2x - 3)$	$(5x + 1)$
Derivative	(2)	(5)

List each factor and the derivative of each.

$$f'(x) = \frac{(\mathbf{2})(5x+1) - (\mathbf{5})(2x-3)}{(5x+1)^2}$$

Multiply each factor by the derivative of the other (bold type). Divide by the square of the denominator.

$$f'(x) = \frac{10x + 2 - 10x + 15}{(5x+1)^2}$$

Multiply in the numerator. (Observe the change of sign in the last two terms.)

$$f'(x) = \frac{17}{(5x+1)^2}$$

Combine like terms for the final answer. It is not necessary to multiply and expand the denominator.

(c) $f(x) = (3x+1)^5 (x^2 - x)^4$ Copy the function.

Factor	$(3x+1)^5$	$(x^2-x)^4$
Derivative	$5(3x+1)^4(3)$	$4(x^2-x)^3(2x-1)$

List each factor and its derivative.

$$f'(x) = [\mathbf{5(3x+1)^4(3)}]\,(x^2-x)^4 + \left[\mathbf{4(x^2-x)^3(2x-1)}\right](3x+1)^5$$

Multiply each factor by the derivative of the other factor. (Each derivative is in boldface type.)

$$f'(x) = 15(3x+1)^4(x^2-x)^4 + (8x-4)(x^2-x)^3(3x+1)^5$$

Simplify the terms. $3 \cdot 5 = 15$, $4(2x-1) = 8x-4$.

$$f'(x) = (3x+1)^4(x^2-x)^3[15(x^2-x) + (8x-4)(3x+1)]$$

Factor out the 4th power of $(3x+1)$ and the 3rd power of (x^2-x).

$$f'(x) = (3x+1)^4(x^2-x)^3[15x^2 - 15x + 24x^2 - 4x - 4]$$

Multiply.

$$f'(x) = (3x+1)^4(x^2-x)^3(39x^2 - 19x - 4)$$

Combine like terms for the final answer.

(d) $f(x) = \dfrac{(2x-5)^4}{(4x-2)^3}$ Copy the function.

	numerator	denominator
Factor	$(2x-5)^4$	$(4x-2)^3$
Derivative	$4(2x-5)^3(2)$	$3(4x-2)^2(4)$

List the numerator and denominator and the derivative of each.

$$f'(x) = \frac{\mathbf{4(2x-5)^3(2)}(4x-2)^3 - (2x-5)^4\mathbf{(3)(4x-2)^2(4)}}{[(4x-2)^3]^2}$$

Multiply each factor by the derivative of the other factor (bold type). Divide by the denominator squared.

$$f'(x) = \frac{4(2x-5)^3(4x-2)^2[2(4x-2)-(2x-5)3]}{(4x-2)^6}$$

Factor the numerator. Multiply the exponents in the denominator.

$$f'(x) = \frac{4(2x-5)^3[2(4x-2)-3(2x-5)]}{(4x-2)^4}$$

Reduce by cancelling $(4x-2)^2$ from the numerator and denominator.

$$f'(x) = \frac{4(2x-5)^3(8x-4-6x+15)}{(4x-2)^4}$$

Multiply. Watch minus signs!

$$f'(x) = \frac{4(2x-5)^3(2x+11)}{(4x-2)^4}$$

Combine like terms for the final answer.

(e) $f(x) = \dfrac{e^x}{\ln x}$

Copy the function.

This solution involves the Quotient Rule.

	numerator	denominator	
factor	e^x	$\ln x$	List the numerator and denominator and their derivatives.
derivative	e^x	$\frac{1}{x}$	

$$f'(x) = \frac{e^x \ln x - e^x\left(\frac{1}{x}\right)}{(\ln x)^2}$$

Multiply each factor by the derivative of the other factor. Divide by the denominator squared.

$$f'(x) = \frac{e^x\left(\ln x - \frac{1}{x}\right)}{(\ln x)^2}$$

Simplify by factoring out e^x.

EXAMPLE 9.2

Write the equation of the tangent line to the graph of $y = \frac{\ln\ (x+2)}{x+2}$ at (–1, 0).

SOLUTION 9.2

We know the slope of the tangent line at (–1, 0) is the first derivative evaluated at $x = -1$. First we'll find $f'(x)$ using the quotient rule.

	numerator	denominator	
factor	$\ln\ (x+2)$	$x+2$	List the numerator and denominator and their derivatives.
derivative	$\frac{1}{x+2}$	1	

$$f'(x) = \frac{(x+2)\frac{1}{x+2} - \ln\ (x+2)\ (1)}{(x+2)^2}$$

Multiply each factor by the derivative of the other factor. Divide by the denominator squared.

$$f'(x) = \frac{1 - \ln\ (x+2)}{(x+2)^2}$$

Simplify.

We need the slope of the tangent when $x = -1$:

$f'(x) = \dfrac{1-\ln(-1+2)}{(-1+2)^2}$ Substitute $x = -1$ into the derivative.

$f'(x) = \dfrac{1-\ln 1}{1} = 1-0 = 1$ The slope of the tangent equals 1.

Now use the point-slope formula:

$y - y_1 = m(x - x_1)$

$y - 0 = 1(x - (-1))$ Substitute $y_1 = 0, x_1 = -1$, and $m = 1$.

$y = x + 1$ This is the equation of the tangent line.

The graph of the function and the tangent line at $x = -1$ follows.

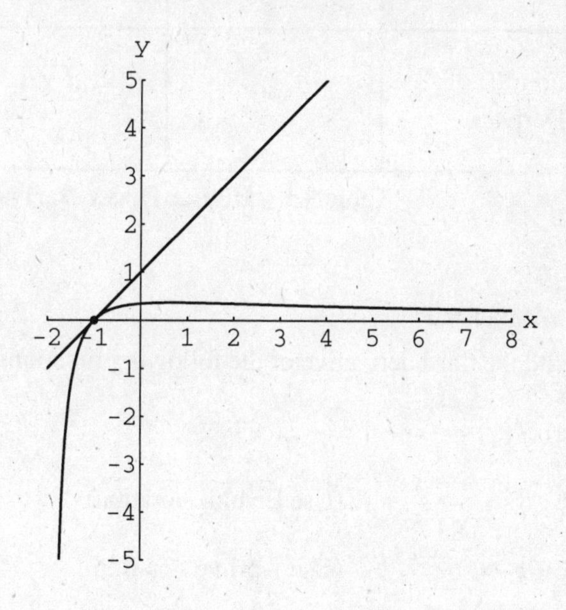

9.2 HIGHER ORDER DERIVATIVES

The differentiation process can be extended from the first derivative, $f'(x)$, to the second, third, etc., by taking the derivative of each successive derivative. The application is very straightforward when we treat each new derivative as an original function. The notation for higher order derivatives is shown in Table 9.1.

Original Function	***y***	***y***	***f(x)***
First Derivative	y'	$\frac{dy}{dx}$	$f'(x)$
Second Derivative	y''	$\frac{d^2y}{dx^2}$	$f''(x)$
Third Derivative	y'''	$\frac{d^3y}{dx^3}$	$f'''(x)$
Fourth Derivative	y''''	$\frac{d^4y}{dx^4}$	$f''''(x)$

Table 9.1 Higher Order Derivatives

EXAMPLE 9.3

Find the third derivative of the following functions.

(a) $f(x) = x^3 - 7x^2 - 2x + 1$

(b) $y = \dfrac{2-x}{x+5}$ (Use Leibniz notation)

(c) $y' = 6e^{-4x}$ (Use *y*-prime notation)

SOLUTION 9.3

(a) $f(x) = x^3 - 7x^2 - 2x + 1$ Copy the function.

$f'(x) = 3x^2 - 14x - 2$ Find the first derivative using the Power Rule.

$f''(x) = 6x - 14$ Find the second derivative.

$f'''(x) = 6$ Find the third derivative.

(b) $y = \dfrac{2-x}{x+5}$ Copy the function.

$\dfrac{dy}{dx} = \dfrac{(-1)(x+5) - (2-x)(1)}{(x+5)^2}$ Use the Quotient Rule to find the first derivative.

$\dfrac{dy}{dx} = \dfrac{-x-5-2+x}{(x+5)^2} = \dfrac{-7}{(x+5)^2}$

$\dfrac{d^2y}{dx^2} = \dfrac{(0)(x+5)^2 - (-7)[2(x+5)^1]}{[(x+5)^2]^2}$

Find the second derivative using the Quotient Rule.

$\dfrac{d^2y}{dx^2} = \dfrac{0 + 7(2)(x+5)}{(x+5)^4}$

$\dfrac{d^2y}{dx^2} = \dfrac{14(x+5)}{(x+5)^4} = \dfrac{14}{(x+5)^3}$

$\dfrac{d^3y}{dx^3} = \dfrac{(0)(x+5)^3 - 14[3(x+5)^2(1)]}{[(x+5)^3]^2}$

Find the third derivative using the Quotient Rule.

$\dfrac{d^3y}{dx^3} = \dfrac{0 - 14(3)(x+5)^2}{(x+5)^6}$

$\dfrac{d^3y}{dx^3} = \dfrac{-42(x+5)^2}{(x+5)^6} = -\dfrac{42}{(x+5)^4}$

Note that when the numerator is a constant, you may find it easier to rewrite the fraction using negative exponents and find the derivative using the Chain Rule. In part two of (b):

$\dfrac{-7}{(x+3)^2} = -7(x+5)^{-2}$

$\dfrac{d^2y}{dx^2} = 14(x+5)^{-3}(1) = \dfrac{14}{(x+5)^3}$ as we found with the Quotient Rule.

(c) $y = 6e^{-4x}$ — Copy the function.

$y' = 6e^{-4x}(-4)$ — Find the first derivative using the Chain Rule for $e^{g(x)}$.

$y' = -24e^{-4x}$

$y'' = -24e^{-4x}(-4)$ — Find the second derivative using the Chain Rule for $e^{g(x)}$.

$y'' = 96e^{-4x}$

$y''' = 96e^{-4x}(-4)$ — Find the third derivative using the Chain Rule for $e^{g(x)}$.

$y''' = -384e^{-4x}$

9.3 IMPLICIT DIFFERENTIATION

The functions we have differentiated in this chapter and chapter 8 have been solved *explicitly* for y, that is, we wrote them in the form $y =$ or $f(x) =$ an expression. However, not all functions will be written (or even *can* be) written explicitly. When a function is written *implicitly*, we treat x and y as functions, and use the Chain Rule to write each derivative of y as y'.

Using the Chain Rule on y

In our first example, we will assume that y represents a function of x. Remember that the derivative of y is y'.

EXAMPLE 9.4

Differentiate with respect to x, assuming that y is a function of x.

(a) x^3

(b) y^3

(c) x^3y^3

SOLUTION 9.4

(a) $D_x[x^3] = 3x^2 \cdot D_x[x]$ — Use the Chain Rule.

$= 3x^2(1)$	$D_x[x] = 1$.
$= 3x^2$	Simplify.

(b) $D_x[y^3] = 3y^2 \cdot D_x[y]$	Use the Chain Rule.
$= 3y^2 \cdot y'$	$D_x[y] = y'$.
$= 3y^2y'$	Simplify.

(c) $D_x[x^3y^3]$ involves a product of x^3 and y^3. We use the Product Rule.

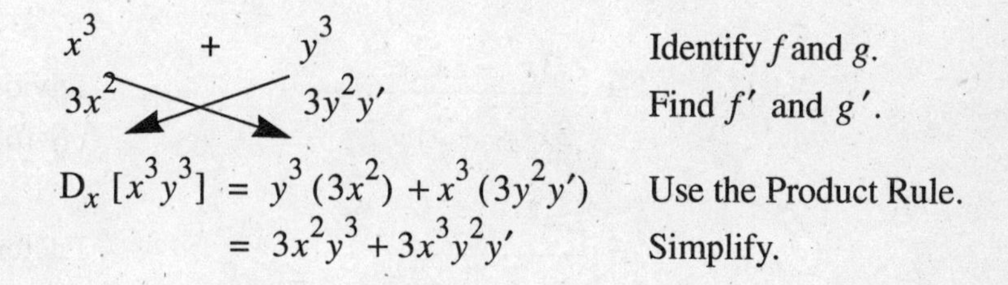

$D_x[x^3y^3] = y^3(3x^2) + x^3(3y^2y')$	Use the Product Rule.
$= 3x^2y^3 + 3x^3y^2y'$	Simplify.

Implicit Differentiation

The following rules will help you find a derivative using implicit differentiation. Keep in mind that the goal is to solve for y'.

> **Finding a First Derivative with Implicit Differentiation**
> 1. Find the derivative of both sides of the equation.
> 2. Collect all terms involving y' on the left side of the equation, and all other terms on the right side of the equation.
> 3. Factor out y' on the left side.
> 4. Divide both sides of the equation by the coefficient of y'.

EXAMPLE 9.5

Find y'.

(a) $x^3 + x^3y^3 + y^3 = 6$

(b) $x^2y^2 - y = 6x$

(c) $e^{xy} = 12x$

(d) $\ln xy = 10x$

SOLUTION 9.5

(a) $D_x[x^3 + x^3y^3 + y^3] = D_x[6]$	Find the derivative of both sides.

$$3x^2 + 3x^2y^3 + 3x^3y^2y' + 3y^2y' = 0$$

See Example 9.4 for the left side. The derivative of a constant is 0.

$$3x^3y^2y' + 3y^2y' = -3x^2 - 3x^2y^3$$

Move all terms with y' to the left, all other terms to the right.

$$y'(3x^3y^2 + 3y^2) = -3x^2 - 3x^2y^3$$

Factor out y'.

$$y' = \frac{-3x^2 - 3x^2y^3}{3x^3y^2 + 3y^2}$$

Divide both sides by the coefficient of y'.

$$y' = \frac{-x^2 - x^2y^3}{x^3y^2 + y^2}$$

Factor out $\frac{3}{3}$ and reduce.

(b) $D_x\ [x^2y^2 - y] = D_x[6x]$

Find the derivative of both sides.

$$y^2(2x) + x^2(2yy') - y' = 6$$

$x^2 + y^2$
$2x \quad 2yy'$

$$2x^2yy' - y' = 6 - 2xy^2$$

Move all terms with y' to the left, all other terms to the right.

$$y'(2x^2y - 1) = 6 - 2xy^2$$

Factor out y'.

$$y' = \frac{6 - 2xy^2}{2x^2y - 1}$$

Divide both sides by the coefficient of y'.

(c) $D_xe^{xy} = D_x(12x)$

Find the derivative of both sides.

$$e^{xy}(1 \cdot y + xy') = 12$$

Use the Chain Rule for e^x. The xy-term requires the Product Rule. The derivative of $12x$ is 12.

$$e^{xy}(y + xy') = 12$$

Simplify inside the parentheses.

$$y + xy' = \frac{12}{e^{xy}}$$

Divide both sides by e^{xy}.

$$xy' = \frac{12}{e^{xy}} - y$$ Subtract y from both sides.

$$y' = \frac{\frac{12}{e^{xy}} - y}{x}$$ Factoring is not necessary. Divide by x.

$$y' = \frac{12 - ye^{xy}}{xe^{xy}}$$ Simplify the fraction.

(d) $D_x \ln xy = D_x 10x$ Take the derivative of both sides.

$$\frac{1}{xy}(1 \cdot y + x \cdot y') = 10$$ Use the Chain Rule for ln x. Remember the Product Rule for the xy-term. The derivative of $10x$ is 10.

$$\frac{1}{xy}(y + xy') = 10$$

$$y + xy' = 10xy$$ Multiply both sides by xy.

$$xy' = 10xy - y$$ Subtract y from both sides.

$$y' = \frac{10xy - y}{x}$$ Divide by x for the final answer.

Slope of a Tangent Line

Recall that the slope of a tangent line at $x = x_0$ to a curve is the first derivative evaluated at $x = x_0$. If the equation of the curve is stated implicitly, you can find the slope of the tangent using implicit differentiation.

EXAMPLE 9.6

Find the slope of the tangent line at the indicated point.

(a) $x^2 + 4y^2 = 100$ at $(-8, 3)$

(b) $4x^2 - 3y^2 + 8x + 16 = 0$ at $(0, \frac{4}{\sqrt{3}})$

SOLUTION 9.6

(a) Since the slope of the tangent line equals the first derivative evaluated at $x = -8$, first find y'.

$D_x[x^2 + 4y^2] = D_x[100]$	Find the derivative of both sides.
$2x + 8yy' = 0$	$D_x[4y^2] = 8yD_x[y]$ $= 8yy'$
$8yy' = -2x$	Isolate y'.
$y' = \frac{-2x}{8y}$	Divide both sides by $8y$.
$y' = -\frac{x}{4y}$	y' is the slope of the tangent at (x, y).

Find the slope at $(-8, 3)$:

$y' = -\frac{(-8)}{4(3)}$	Substitute $x = -8$ and $y = 3$.
$y' = \frac{2}{3}$	Reduce.

(b) $D_x[4x^2 - 3y^2 + 8x + 16] = D_x[0]$	Find the derivative of both sides.
$8x - 6yy' + 8 = 0$	$D_x[-3y^2] = -6y \cdot D_x[y]$ $= -6yy'$.
$-6yy' = -8x - 8$	Isolate y'.
$y' = \frac{-8x - 8}{-6y}$	Divide both sides by $-6y$.
$y' = \frac{4x + 4}{3y}$	y' is the slope of the tangent at (x, y).

Find the slope at $(0, \frac{4}{\sqrt{3}})$:

$y' = \frac{4(0) + 4}{3\left(\frac{4}{\sqrt{3}}\right)}$	Substitute $x = 0$, $y = \frac{4}{\sqrt{3}}$.

$$y' = \frac{4}{\frac{12}{\sqrt{3}}} = 4 \cdot \frac{\sqrt{3}}{12} = \frac{\sqrt{3}}{3}$$ Simplify.

EXAMPLE 9.7

Find the equation of the tangent line at the indicated point.

(a) $x^2 + 4y^2 = 100$ at $(-8, 3)$

(b) $4x^2 - 3y^2 + 8x + 16 = 0$ at $(0, \frac{4}{\sqrt{3}})$

SOLUTION 9.7

(a) We have already found the slope of the tangent at $(-8, 3)$ in Example 9.6(a).
Use the point-slope formula to find the equation of the tangent line.

$$y - y_1 = m(x - x_1)$$ Write the point-slope formula.

$$y - 3 = \frac{2}{3}(x - (-8))$$ Substitute $y_1 = 3$, $m = \frac{2}{3}$, $x_1 = -8$.

$$y - 3 = \frac{2}{3}(x + 8)$$ Simplify.

$$y = \frac{2}{3}x + \frac{25}{3}$$ $\frac{2}{3} \cdot 8 + 3 = \frac{16}{3} + \frac{9}{3} = \frac{25}{3}$

(b) $$y - y_1 = m(x - x_1)$$ Write the point-slope formula.

$$y - \frac{4}{\sqrt{3}} = \frac{\sqrt{3}}{3}(x - 0)$$ Substitute $y_1 = \frac{4}{\sqrt{3}}$, $m = \frac{\sqrt{3}}{3}$, $x_1 = 0$.

$$y = \frac{\sqrt{3}}{3}x + \frac{4}{\sqrt{3}}$$ Solve for y.

or

$$y = \frac{\sqrt{3}x + 4\sqrt{3}}{3}$$ Answer may be written in this form. Note that $\frac{4}{\sqrt{3}} = \frac{4}{\sqrt{3}} \cdot \frac{\sqrt{3}}{\sqrt{3}} = \frac{4\sqrt{3}}{3}$

Practice Exercises

1. Find the derivative of the following functions using the product and quotient rules.

   (a) $f(x) = (x^2 - 3)(5x^3 + 2x)$

   (b) $f(x) = \dfrac{(2x-5)}{(7x+3)}$

   (c) $f(x) = (x-1)^7 (4x^2 - x)^5$

   (d) $f(x) = \left(\dfrac{4x-3}{x^2+5}\right)^2$

   (e) $f(x) = (3x-2)^{1/2}(x+6)^{1/4}$

   (f) $f(x) = \sqrt{\dfrac{6x+1}{5x-3}}$

   (g) $f(x) = \dfrac{e^x + 1}{\ln x}$

   (h) $f(x) = (e^x + 2)(x^2 + 3x - 1)$

2. Find the third derivative of the following functions.

   (a) $f(x) = x^4 - 2x^3 + 3x^2 - x$

   (b) $f(x) = \dfrac{x}{x-1}$

   (c) $y = 4\ln x^2$

   (d) $y = 5e^x$

3. Find the following derivatives through implicit differentiation.

   (a) $4y - xy = 6$

   (b) $\dfrac{x}{y} = 12x^2$

   (c) $\sqrt{y} + \sqrt{x} = 5$

   (d) $\ln(xy) = x^2 - 1$

   (e) $e^{\sqrt{y}} = x$

   (f) $x^3 y + y^2 = 14$

4. Find the equation of the tangent line at the indicated point. Write your answers in $y = mx + b$ form.

   (a) $y = \dfrac{\ln(x+1)}{(x+1)}$ at $(0, 0)$

   (b) $4x^2 + y^2 = 100$ at $(3, -8)$

   (c) $9x^2 - 16y^2 + 192y - 720 = 0$ at $(4\sqrt{5}, 0)$

Answers

1. (a) $25x^4 - 39x^2 - 6$

(b) $\dfrac{41}{(7x+3)^2}$

(c) $(x-1)^6(4x^2-x)^4(68x^2-52x+5)$

(d) $\dfrac{-4(4x-3)(2x^2-3x-10)}{(x^2+5)^3}$

(e) $\dfrac{9x+34}{4(x+6)^{3/4}(3x-2)^{1/2}}$

(f) $\dfrac{-23}{2(5x-3)^{3/2}(6x+1)^{1/2}}$

(g) $\dfrac{xe^x \ln x - e^x - 1}{x(\ln x)^2}$

(h) $x^2e^x + 5xe^x + 2e^x + 4x + 6$

2. (a) $24x - 12$

(b) $\dfrac{-6}{(x-1)^4}$

(c) $\dfrac{16}{x^3}$

3. (a) $\dfrac{y}{4-x}$

(b) $\dfrac{y - 24xy^2}{x}$

(c) $\dfrac{-\sqrt{y}}{\sqrt{x}}$

(d) $\dfrac{2x^2y - y}{x}$

(e) $\dfrac{2\sqrt{y}}{e^{\sqrt{y}}}$

(f) $\dfrac{-3x^2y}{x^3+2y}$

4. (a) $y = x$

(b) $y = \dfrac{3}{2}x - \dfrac{25}{2}$

(c) $y = \dfrac{-9\sqrt{5}}{24}x + \dfrac{15}{2}$

10

Applications of Derivatives

In this chapter we will use derivatives to solve word problems and to analyze features of graphs. We will solve related rates problems, that is, problems in which related variables change over time, and maximum and minimum problems.

10.1 RELATED RATES WORD PROBLEMS

Related rates problems involve one or more variables that change over time. Since the derivatives in these problems will be taken with respect to time, rather than with respect to x, we will use implicit differentiation (see section 9.3). For example, the statement

"the radius increases at a rate of 5 feet per second"

would be translated symbolically as

$$\frac{dr}{dt} = 5 \text{ feet/sec}$$

where r is the length of the radius.

EXAMPLE 10.1

Translate each statement into symbols.

(a) The radius of a sphere is increasing at a rate of 10 feet per second.

(b) The height of a cone is decreasing at a rate 6 inches per second.

SOLUTION 10.1

(a) Let r = radius of the sphere.

Then $\frac{dr}{dt}$ = the rate of change of the radius.

The statement translates as $\frac{dr}{dt} = 10$ feet/sec.

(b) Let h = height of the cone.

Then $\frac{dh}{dt}$ = the rate of change of the height.

The statement translates as $\frac{dh}{dt} = -6$ in/sec, where the negative sign indicates a decrease.

Implicit Differentiation and Related Rates

Before solving some word problems, let's practice implicit differentiation on some of the formulas we will encounter in the word problems.

EXAMPLE 10.2

Differentiate each equation with respect to t.

(a) $A = \pi r^2$

(b) $V = \pi r^3$

(c) $V = \frac{1}{3}\pi r^2 h$

SOLUTION 10.2

(a) $A = \pi r^2$ — Given equation.

$\frac{dA}{dt} = \pi(2r^1)\frac{dr}{dt}$ — Find the derivative of each side.

$\frac{dA}{dt} = 2\pi r\frac{dr}{dt}$ — Simplify.

(b) $V = \pi r^3$ — Given equation.

$\frac{dV}{dt} = \pi(3r^2)\frac{dr}{dt}$ — Find the derivative of each side.

$\frac{dV}{dt} = 3\pi r^2\frac{dr}{dt}$ — Simplify.

(c) $V = \frac{1}{3}\pi r^2 h$ Given equation.

$$\frac{dV}{dt} = \frac{1}{3}\pi\left[h2r\frac{dr}{dt} + r^2\frac{dh}{dt}\right]$$

Use the Product Rule since both h and r are functions of t.

$$\frac{dV}{dt} = \frac{1}{3}\pi\left[2hr\frac{dr}{dt} + r^2\frac{dh}{dt}\right]$$

Simplify.

Solving Related Rates Problems

Some general guidelines for setting up related rates problems are:

1. Choose symbols to represent the variables.
2. Find an equation that describes the relationship between the variables.
3. Express the given information symbolically, taking care to represent each rate as a derivative with respect to time. Identify what you are looking for using symbols.
4. Differentiate both sides of the equation from step 1 with respect to time.
5. Substitute given information to find whatever is required.

EXAMPLE 10.3

Solve each related rates problem.

(a) The price of a stock varies according to the equation

$$p(t) = 10 + 2\sqrt{t}$$

where: p is in dollars per share
t is time in months from the stock's first offering.
At what rate is the price increasing at the 4th month after the stock is first offered?

(b) City planners have determined that a city is growing as given by the equation

$$P(t) = 10{,}000\,(e^{0.2t})$$

where P is the population and t is the time in years after 1980. Find the rate of increase in population in 1985.

(c) The concentration of a certain drug in the bloodstream t minutes after it is injected is:

$$C(t) = 10(\ln t) - t^2$$

where C is the concentration in micrograms/deciliter. Find the change in concentration 3 minutes after injection.

(d) An unfortunate accident has caused a rupture in the hull of a petroleum tanker. The resulting oil slick is spreading in a circular manner. If the radius of this circle is increasing at 0.2 kilometers per hour, how fast is the area covered by the oil slick expanding when the radius is 1 kilometer?

(e) A spherically shaped star is uniformly expanding at the rate such that its radius is increasing at 100 kilometers per second. How fast is the volume of the star increasing when its diameter is 400,000 kilometers? Use

$$V = \frac{4}{3}\pi r^3$$

where r is the radius and V is the volume.

(f) A youthful observer watches his toy rocket rise at the rate of 20 feet/second. The youth stands 300 feet from the rocket launchpad as he watches the rocket rise. How fast is the line-of-sight distance increasing between him and the rocket when the rocket is 400 feet in the air? Assume the flight path is perfectly vertical.

SOLUTION 10.3

(a) $$p(t) = 10 + 2\sqrt{t} = 10 + 2t^{1/2}$$

$$\frac{dp}{dt} = p'(t) = 2\left(\frac{1}{2}\right)t^{-1/2} = t^{-1/2}$$

The rate of change in price is the first derivative of the price equation.

$$p'(4) = \frac{1}{\sqrt{4}} = \frac{1}{2} \text{ dollars/month}$$

The rate of change in price at the 4th month after first issue is the first derivative evaluated at $t = 4$.

(b) $$P(t) = 10{,}000\,(e^{0.2t})$$

$$P'(t) = 10{,}000\,(e^{0.2t})\,(0.2)$$

$$= 2000e^{0.2t}$$

The change in population is the first derivative of the population equation.

$$\begin{aligned} P'(5) &= 2000e^{0.2(5)} \\ &= 2000e^{1} \\ &\approx 2000\,(2.718) \\ &= 5436 \text{ people/year} \end{aligned}$$

The change in population per year at the 5th year after 1980 (1985) is the derivative evaluated at $t = 5$.

(c) $C(t) = 10\ln t - t^2$

$C'(t) = 10\left(\frac{1}{t}\right) - 2t$

$= \frac{10}{t} - 2t$

The change in concentration is the first derivative of the concentration equation.

$C'(3) = \frac{10}{3} - 2(3)$

$= 3.33 - 6$

$= -2.67 \frac{\text{micrograms}}{\text{deciliter}}$

The change in concentration 3 minutes after injection is the first derivative evaluated at $t = 3$.

The negative sign means that the concentration is decreasing over time.

(d)

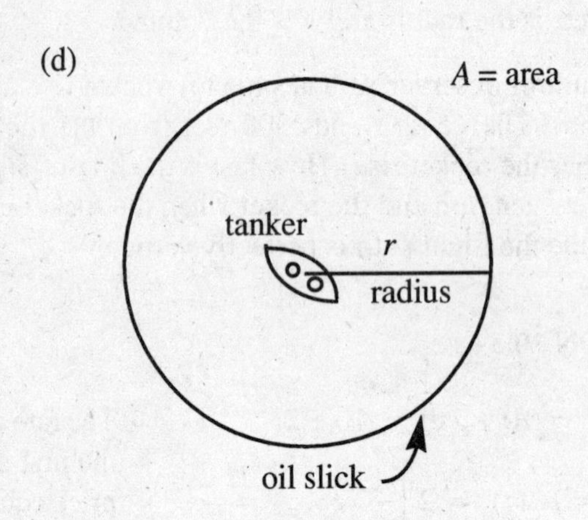

Draw a sketch and select and label appropriate variables.

A = area

$\frac{dr}{dt}$ = rate of change in radius = 0.2 km/hr

$\frac{dA}{dt}$ = rate of change in area = ?

r = radius at desired moment = 1 km

The known rates and variable values are listed.

$A = \pi r^2$

Write an equation that relates the variables.

$\frac{dA}{dt} = 2\pi r \frac{dr}{dt}$

Find the derivative with respect to time.

$$\frac{dA}{dt} = 2\pi(1 \text{ km})\left(0.2 \frac{\text{km}}{\text{hr}}\right)$$

Substitute the given information.

$$\frac{dA}{dt} = 2\pi(1)(0.2)$$
$$= 0.4\pi \approx 1.257$$

All values needed are known. Multiply to find the answer.

The rate of change in area, at the instant the radius is 1 kilometer, is 1.257 sq. km/hr.

(e) V = Volume

Draw a sketch and select and label appropriate variables.

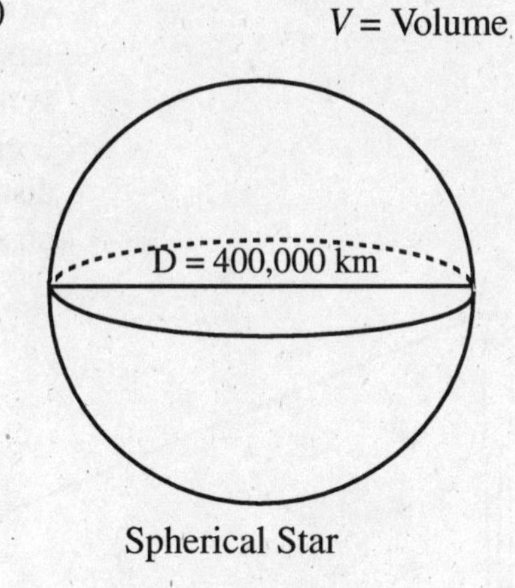

Spherical Star

V = volume

The known rates and values are listed.

$\frac{dr}{dt}$ = change in radius = 100 km/sec

$\frac{dV}{dt}$ = change in volume = ?

$d = 400{,}000 \text{ km} = 2r$

$r = 200{,}000 \text{ km}$ = radius

$$V = \frac{4}{3}\pi r^3$$

Write an equation that relates the volume and the radius.

$$\frac{dV}{dt} = \frac{4}{3}\pi 3r^2\frac{dr}{dt}$$
$$= 4\pi r^2\frac{dr}{dt}$$

Find the derivative using the Chain Rule.

$$\frac{dV}{dt} = 4\pi\,(200{,}000 \text{ km})^2\left(100 \ \frac{\text{km}}{\text{sec}}\right)$$ Substitute the given values.

$$\frac{dV}{dt} = \text{change in volume}$$

$$= 4\,(\pi)\,(200{,}000\,)^2\,(100)$$ Multiply the values to find the answer.

$$= 16{,}000{,}000{,}000{,}000\ \pi \text{ cubic km/sec}$$

(f) Construct a diagram and label it. We choose to label vertical distance as y, horizontal distance as x, distance from rocket to observer as s.

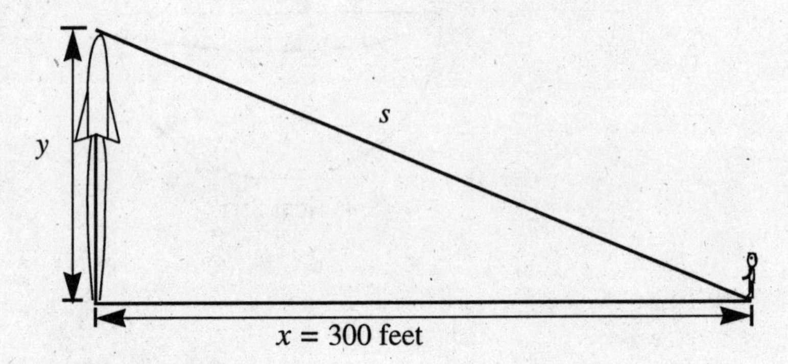

x = distance from launchpad to observer = 300 ft
y = height of rocket = 400 ft
s = line of sight distance from observer to rocket

List all the known rates and values.

$$\frac{ds}{dt} = \text{change in line-of-sight distance} = ?$$

$$\frac{dy}{dt} = 20 \text{ ft/sec}$$

$$x^2 + y^2 = s^2$$

Write an equation that relates the variables. This equation follows from the Pythagorean Theorem for a right triangle.

$$2x\frac{dx}{dt} + 2y\frac{dy}{dt} = 2s\frac{ds}{dt}$$

Use implicit differentiation with respect to time.

$$x\frac{dx}{dt} + y\frac{dy}{dt} = s\frac{ds}{dt}$$

Divide by 2.

$$(300)\left(\frac{dx}{dt}\right) + (400)\ (20) = s\left(\frac{ds}{dt}\right)$$

Substitute.

$$\frac{dx}{dt} = 0$$

We can see that we are missing two derivative values and a value for s. However, because the distance from the launchpad to the observer does not change, the derivative of x with respect to time is 0.
We find s using

$$x^2 + y^2 = s^2$$

$$x^2 + y^2 = s^2$$

$$300^2 + 400^2 = s^2$$

Substitute known values.

$$9000 + 16{,}000 = s^2$$

Square each number.

$$25{,}000 = s^2$$

Add.

$$500 = s$$

Take the square root of both sides.

$$300\ (0) + 400\ (20) = 500\left(\frac{ds}{dt}\right)$$

Now the value of $\frac{ds}{dt}$ can be found.

$$0 + 8000 = 500\left(\frac{ds}{dt}\right)$$

$$\frac{8000}{500} = \frac{ds}{dt}$$

$$16 \text{ ft/sec} = \frac{ds}{dt}$$

10.2 FEATURES OF GRAPHS

In this section we will use graphs to introduce the concepts of increasing and decreasing intervals, concave up and concave down intervals, and inflection points. The next two sections will link calculus to these concepts.

Increasing and Decreasing Intervals

A function is increasing when its y-values increase as its x-values increase. A function is decreasing when its y-values decrease as its x-values increase. Observe that $f(x) = x^2$ decreases when $x < 0$ and increases when $x > 0$:

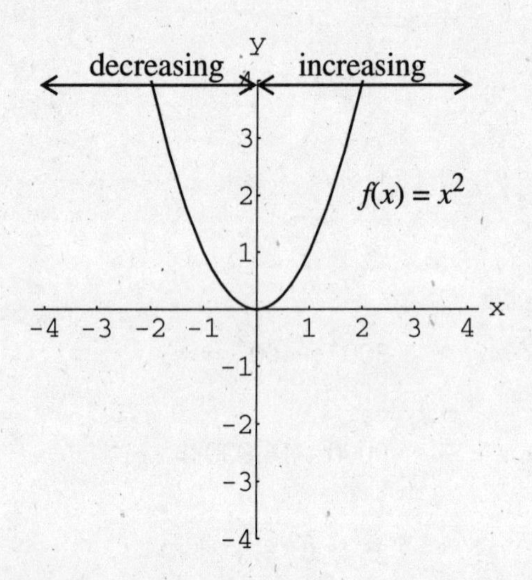

Using interval notation, $f(x)$ is decreasing on $(-\infty, 0)$ and increasing on $(0, \infty)$.

EXAMPLE 10.4

Use the given sketch to determine the intervals where $f(x)$ is increasing and where $f(x)$ is decreasing.

SOLUTION 10.4

The function is increasing from $(-\infty, -3)$ and from $(2, \infty)$. The function is decreasing from $(-3, 2)$. Note from this example that the intervals are given in terms of x values.

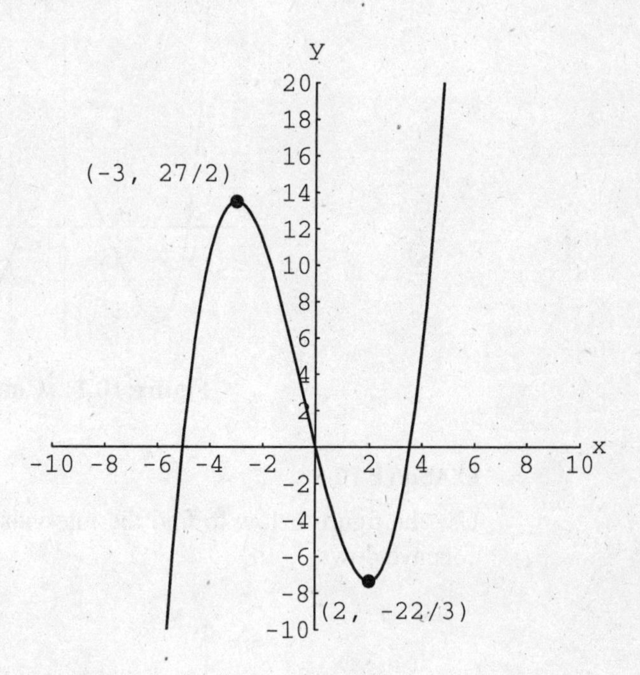

Extrema

Extrema of a function on its entire domain are the maximum and/or minimum y-values of the function (also called the absolute maximum and absolute minimum). We will also be interested in finding the relative extrema of a function—that is, the maximum or minimum value within a subinterval of its domain. Looking at the graph used for Example 10.4, we call $(-3, 27/2)$ a relative maximum and $(2, -22/3)$ a relative minimum. This graph has no absolute maximum because the curve continues to rise on the right. The graph has no absolute minimum because the curve continues to fall on the left.

Concavity

When the graph of a function $f(x)$ lies above its tangent lines, $f(x)$ is concave upward. When the graph of $f(x)$ lies below its tangent lines, $f(x)$ is concave downward. It may also help to think of the graph as "holding water" (concave up) or "pouring water" (concave down). In Figure 10.1, $f(x)$ is concave up from $(-\infty, a)$ and from (b, ∞) and is concave down on (a, b).

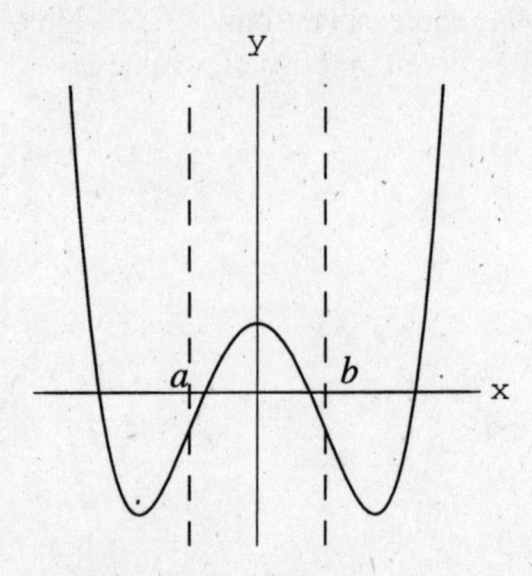

Figure 10.1 Concavity

EXAMPLE 10.5

Use the figure below to find the intervals where $f(x)$ is concave up and concave down.

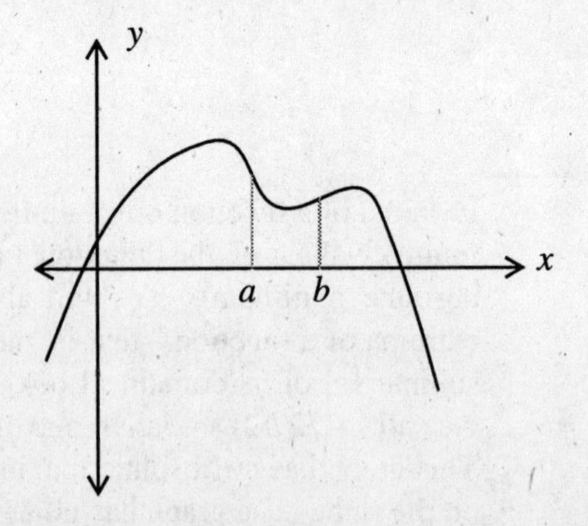

SOLUTION 10.5

The function is concave up on (a, b).
The function is concave down on $(-\infty, a)$ and (b, ∞).

Inflection Points

If there is a point where a function changes concavity, it is called an *inflection point*. Figure 10.1 has two points of inflection, $(a, f(a))$ and $(b, f(b))$.

EXAMPLE 10.6

Use the graph below to find the inflection points.

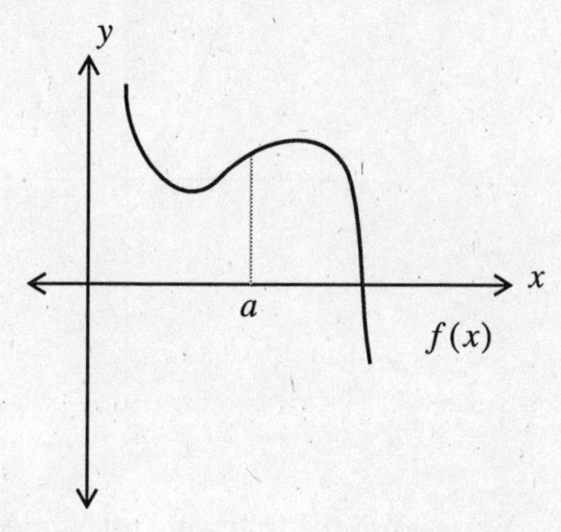

SOLUTION 10.6

The graph has one inflection point, $(a, f(a))$.

Note that in this section we used graphs to *approximate* the intervals and the inflection points. In the next section we will use calculus to find *exact* x values for increasing and decreasing intervals, concave up and down intervals, and exact x and y values for inflection points.

10.3 FIRST AND SECOND DERIVATIVE TESTS AND CONCAVITY

Increasing and Decreasing Intervals

By sketching some tangent lines where $f(x) = x^2$ is decreasing and increasing, we note that the function decreases when the slopes of the tangents are negative, and increases when the slopes of the tangents are positive:

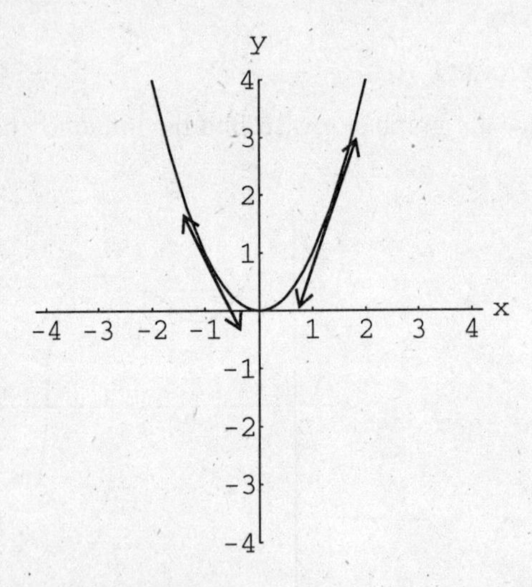

We can use the following rules to determine when a differentiable function is increasing or decreasing.

To Determine Intervals Where $f(x)$ Is Increasing or Decreasing:

1. Find $f'(x)$.
2. Find critical numbers—that is, values of x for which $f'(x) = 0$ or where $f'(x)$ is undefined.
3. Draw a number line, dot in the critical numbers. Determine the sign of $f'(x)$ within each region by choosing any number within the region and substituting it into $f'(x)$. You cannot use a critical number to determine the sign.
4. Where $f'(x) > 0$ (+ on the number line), $f(x)$ is increasing.
5. Where $f'(x) < 0$ (– on the number line), $f(x)$ is decreasing.

EXAMPLE 10.7

Find the intervals on which $f(x) = x^3 - 4x^2 + 1$ is increasing or decreasing.

SOLUTION 10.7

$f'(x) = 3x^2 - 8x$ Find $f'(x)$.

$3x^2 - 8x = 0$ Set $f'(x) = 0$.

$x(3x - 8) = 0$ Factor.

$x = 0$ or $3x - 8 = 0$, $x = \frac{8}{3}$ Solve for critical numbers.

$f'(x)$: + (left of 0), − (between 0 and 8/3), + (right of 8/3)

$f'(-1) = +11$ so mark + left of 0. $f'(1) = -5$ so mark – between 0 and 8/3. $f'(5) = +3$ so mark + right of 8/3.

$f(x)$ is increasing on $(-\infty, 0)$ and $(\frac{8}{3}, \infty)$.

$f(x)$ is decreasing on $(0, \frac{8}{3})$.

EXAMPLE 10.8

Find the intervals on which $f(x) = \frac{x^2}{x^2 - 4}$ is increasing and decreasing.

SOLUTION 10.8

$$f'(x) = \frac{(x^2 - 4)(2x) - (x^2)(2x)}{(x^2 - 4)^2}$$ Use the Quotient Rule.

$$f'(x) = \frac{-8x}{(x^2 - 4)^2}$$ Simplify.

Set the numerator equal to 0 to find where $f'(x) = 0$.

$-8x = 0$ Solve for x.

$x = 0$ $x = 0$ is a critical number.

Set the denominator equal to 0 to find where $f'(x)$ is undefined.

$(x^2 - 4)^2 = 0$ Set the denominator = 0.

$x^2 - 4 = 0$ Take the square root of both sides.

$x^2 = 4$ Add 4 to both sides.

$x = \pm 2$ Take the square root of both sides.

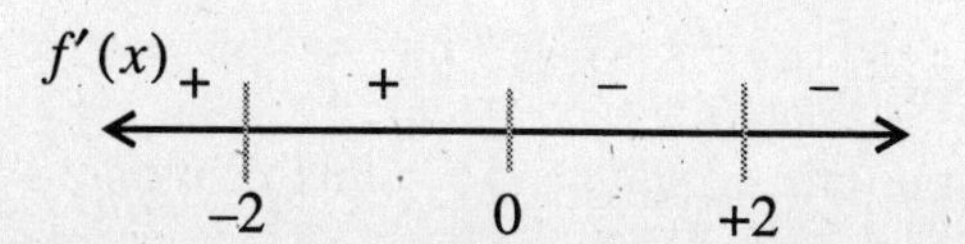

Since the denominator is always + when $x \neq \pm 2$, check the sign of $-8x$.

$f(x)$ is increasing on $(-\infty, -2)$ and $(-2, 0)$.
$f(x)$ is decreasing on $(0, 2)$ and $(2, \infty)$.

Local Extrema

We have already noted that in intervals where a function is decreasing, tangents to the curve have negative slopes, and where a function is increasing, tangents to the curve have positive slopes. Where the slope of the tangent changes from positive to negative or negative to positive, we have a local maximum or local minimum to the curve. The First Derivative Test states that if the signs of the first derivative are different on either side of a critical number, c, then $f(c)$ is a relative maximum or minimum. The following steps apply when f is a continuous function on an open interval, differentiable except possibly at a critical number.

1. Find $f'(x)$.
2. Find critical numbers—that is, all numbers c such that $f'(c) = 0$ or $f'(c)$ is undefined.
3. Draw a number line, dot in the critical numbers and determine the sign of $f'(x)$ in each region.
4. Use the following guides to determine whether $f(c)$ is a relative maximum or minimum.

sign of $f'(x)$ — $f(c)$ is a relative maximum

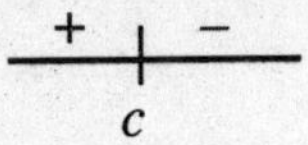

+ | −
c

sign of $f'(x)$ — $f(c)$ is a relative minimum

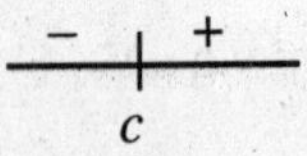

− | +
c

sign of $f'(x)$ — $f(c)$ is neither a relative maximum or minimum.

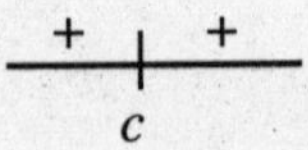

− | − + | +
c c

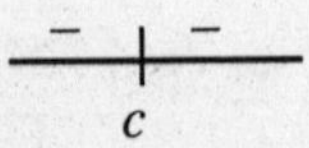

EXAMPLE 10.9

Find the relative extrema for $f(x) = x^3 - 4x^2 + 1$.

SOLUTION 10.9

We found the critical numbers for this continuous function in Example 10.7.

We found the following sign changes for $f'(x)$.

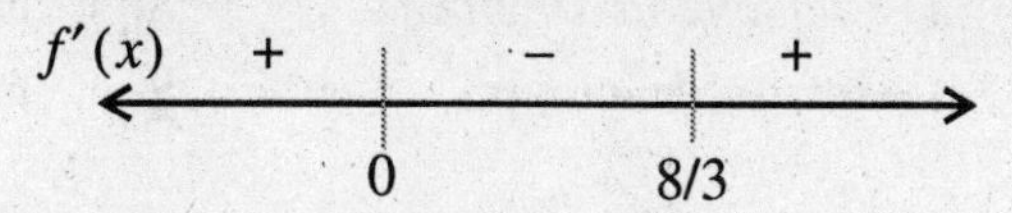

The First Derivative Test tells us that:

$x = 0$ is a relative maximum	Sign change + to –.
$x = \frac{8}{3}$ is a relative minimum	Sign change – to +.

$f(0) = 1$, so 1 is a relative maximum that occurs at $x = 0$.

$f(\frac{8}{3}) = -\frac{229}{27}$ so $-\frac{229}{27}$ is a relative minimum that occurs at $x = 8/3$.

EXAMPLE 10.10

Find the open intervals (if any) on which $f(x) = 3x^4 - x^3 - 24x^2 + 12x$ is increasing or decreasing and find all relative extrema.

SOLUTION 10.10

$f'(x) = 12x^3 - 3x^2 - 48x + 12$	Find $f'(x)$.
$12x^3 - 3x^2 - 48x + 12 = 0$	Set $f'(x) = 0$ to find critical numbers.
$3x^2(4x - 1) - 12(4x - 1) = 0$	Factor by grouping.
$(4x - 1)(3x^2 - 12) = 0$	Factor out $4x - 1$.
$4x - 1 = 0$ or $3x^2 - 12 = 0$ $x = \frac{1}{4}$ $\quad x = \pm 2$	Set each factor equal to 0 and solve.

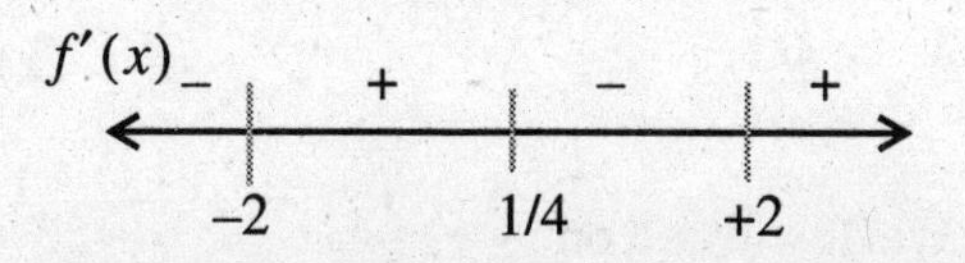

Determine the sign of $f'(x)$ in each region.

$f(x)$ is increasing on $(-2, \frac{1}{4})$ and $(2, \infty)$.

$f(x)$ is decreasing on $(-\infty, -2)$ and $(\frac{1}{4}, 2)$.

$f(-2) = -64$ is a relative minimum	Sign changes – to +.
$f(\frac{1}{4}) = \frac{383}{256}$ is a relative maximum	Sign changes + to –.
$f(2) = -32$ is a relative minimum	Sign changes – to +.

Concavity

It can be shown that a function is concave up when $f''(x) > 0$ and concave down when $f''(x) < 0$, we can use the following steps to determine a differentiable function's concavity.

To Find Intervals of Concavity

1. Find $f''(x)$.
2. Find x-values where $f''(x) = 0$ and where $f''(x)$ is undefined.
3. Draw a number line, dot in the values found in step 2 and determine the sign of $f''(x)$ in each region thus formed.
4. The function is concave upward where $f''(x) > 0$ and concave downward where $f''(x) < 0$.

EXAMPLE 10.11

Determine the intervals where $f(x) = x^3 - 4x^2 + 1$ is concave upward and downward.

SOLUTION 10.11

$f'(x) = 3x^2 - 8x$	Find $f'(x)$.
$f''(x) = 6x - 8$	Find $f''(x)$.
$6x - 8 = 0$	Set $f''(x) = 0$.
$x = \frac{8}{6} = \frac{4}{3}$	Solve for x.
$f''(x)$: – to the left of 4/3, + to the right of 4/3 (number line)	Dot in $x = 4/3$. Find the sign of $f''(x)$ in each region.

$f(x)$ is concave downward on $(-\infty, \frac{4}{3})$.

$f(x)$ is concave upward on $(\frac{4}{3}, \infty)$.

EXAMPLE 10.12

Determine the intervals where $f(x) = \frac{1}{x^2 - 4}$ is concave upward or downward.

SOLUTION 10.12

$f(x) = (x^2 - 4)^{-1}$ Rewrite $f(x)$.

$f'(x) = -1(x^2 - 4)^{-2}(2x)$ Use the Chain Rule.

$f'(x) = -2x(x^2 - 4)^{-2}$ Simplify.

$f''(x) = (x^2 - 4)^{-2}(-2) + (-2x)(-4x(x^2 - 4)^{-3})$

Use the Product Rule.

$f''(x) = -2(x^2 - 4)^{-3}[(x^2 - 4) + x(-4x)]$

Factor.

$f''(x) = -2(x^2 - 4)^{-3}(-3x^2 - 4)$ Simplify.

$f''(x) = \frac{2(3x^2 + 4)}{(x^2 - 4)^3}$ Simplify.

$f''(x) = 0$ when the numerator equals 0.

$2(3x^2 + 4) = 0$ Find $f''(x) = 0$ by setting the numerator of $f'' = 0$.

$3x^2 + 4 = 0$ Solve for x.

$x^2 = -\frac{4}{3}$ This has no real solutions so there are no real values of x where $f''(x) = 0$.

$f''(x)$ is undefined when the denominator equals 0.

$(x^2 - 4)^3 = 0$ — Set the denominator equal to 0.

$x^2 - 4 = 0$ — Solve for x.

$x = \pm 2$

$f''(x)$ + | − | + — Determine the sign of $f''(x)$ in each region.

−2 +2

$f(x)$ is concave upward on $(-\infty, -2)$ and $(2, \infty)$.

$f(x)$ is concave downward on $(-2, 2)$.

Inflection Points

The point where a function changes from being concave upward to concave downward or vice versa is called an inflection point. The procedure for finding inflection points is the same as the procedure for finding intervals of concavity. Be careful to determine whether the x-values where the concavity changes are in the domain of $f(x)$.

EXAMPLE 10.13

Find the inflection points, if any, for each function.

(a) $f(x) = x^3 - 4x^2 + 1$

(b) $f(x) = \dfrac{1}{x^2 - 4}$

SOLUTION 10.13

(a) Returning to Example 10.11, we know the sign of $f''(x)$

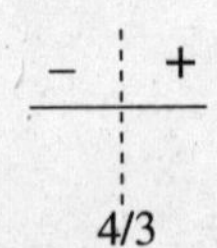

Thus the concavity changes at $x = \frac{4}{3}$. The inflection point is

$$\left(\frac{4}{3}, f\left(\frac{4}{3}\right)\right) = \left(\frac{4}{3}, -\frac{101}{27}\right).$$

(b) Returning to Example 10.12, we know the

sign of $f''(x)$

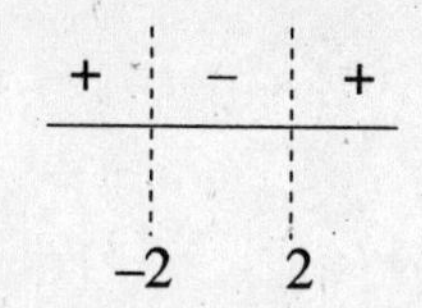

However, $x = -2$ and $x = 2$ are *not* inflection points because the denominator of $f(x) = 1/(x^2 - 4)$ does *not* include −2 and 2. There are no inflection points for $f(x)$.

Second Derivative Test

An alternative technique exists for finding relative maximums and minimums. This technique is called the Second Derivative Test.

To Find Relative Extrema Using the Second Derivative Test:

1. Find $f'(x)$.
2. Find $f''(x)$.
3. Solve $f'(c) = 0$.
4. If $f''(c) > 0$, then $f(c)$ is a relative minimum.
   If $f''(c) < 0$, then $f(c)$ is a relative maximum.
   If $f''(c) = 0$, return to the First Derivative Test to analyze $f(c)$ and watch for points of inflection which can occur when $f''(c) = 0$.

EXAMPLE 10.14

Use the Second Derivative Test to find the relative extrema for each function.

(a) $f(x) = 2x^4 + 3x^3 - 1$

(b) $f(x) = x^5 - x^4 + 1$

SOLUTION 1

(a) $f'(x) = 8x^3 + 9x^2$ Find $f'(x)$.

$f''(x) = 24x^2 + 18x$ Find $f''(x)$.

$8x^3 + 9x^2 = 0$ Set $f'(x) = 0$.

$x^2(8x + 9) = 0$ Factor.

$x = 0 \quad \text{or} \quad x = -\dfrac{9}{8}$ Solve.

$f''(0) = 24(0)^2 + 18(0) = 0$ Find $f''(0)$.

$f''\left(-\dfrac{9}{8}\right) = 24\left(-\dfrac{9}{8}\right)^2 + 18\left(-\dfrac{9}{8}\right) = \dfrac{81}{8}$ Find $f''\left(-\dfrac{9}{8}\right)$.

Since $f''(0) = 0$, the Second Derivative Test fails.

First Derivative Test:

+ ⋮ + sign of $f'(x)$

0

Since there is no sign change, (0, –1) is neither a relative maximum nor a relative minimum. It is, however, an inflection point:

– ⋮ + sign of $f''(x)$

0

Since $f''\left(-\dfrac{9}{8}\right) > 0$, $\left(-\dfrac{9}{8}, -2.06\right)$ is a relative minimum.

Here is a sketch of $f(x) = 2x^4 + 3x^3 - 1$. Note the relative minimum at $x = -9/8$ and the inflection point at $x = 0$.

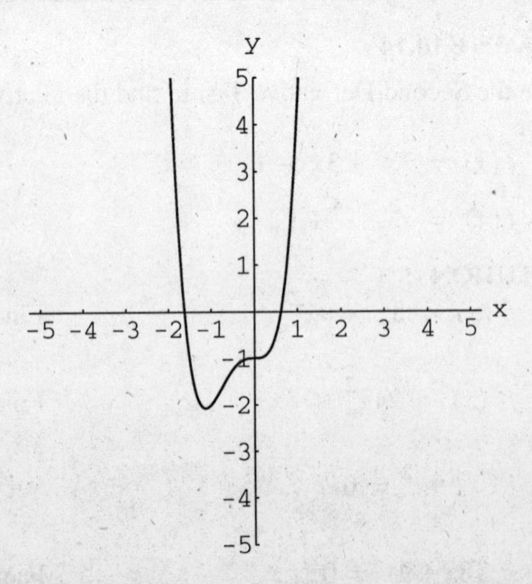

(b) $f(x) = x^5 - x^4 + 1$

$$f'(x) = 5x^4 - 4x^3$$

$$f''(x) = 20x^3 - 12x^2$$

$$5x^4 - 4x^3 = 0$$

$$x^3(5x - 4) = 0$$

$$x = 0 \quad \text{or} \quad x = \frac{4}{5}$$

$$f''(0) = 20(0)^3 - 12(0)^2 = 0 \qquad \text{Test fails.}$$

$$f''\left(\frac{4}{5}\right) = 20\left(\frac{4}{5}\right)^3 - 12\left(\frac{4}{5}\right)^2 = 2\frac{14}{25} \quad f''\left(\frac{4}{5}\right) > 0.$$

We know $\left(\frac{4}{5}, \frac{2869}{3125}\right)$ is a relative minimum. Since $f''(0) = 0$, we use the First Derivative Test:

sign of $f'(x)$

$$\begin{array}{c|c} + & + \\ \hline & \\ \end{array}$$
$$0$$

Since there is no sign change, (0, 0) is neither a relative maximum nor a relative minimum (try testing for an inflection point).

10.4 CURVE SKETCHING USING THE TOOLS OF CALCULUS

We can now put all our techniques for analyzing graphs of functions together to draw a sophisticated graph.

Graphing Using Calculus

We can summarize the various graphing aids which can be used to sketch a function. This list is not meant to be memorized, but to act as a guide to help you prepare a sophisticated graph of a function.

1. Find the domain of $f(x)$ and, if possible, the range of f.

2. Find any x-intercept(s) and the y-intercept.
3. Dot in vertical asymptotes.
4. Find relative extrema.
5. Find intervals where $f(x)$ is increasing and decreasing (use the first derivative) and intervals where $f(x)$ is concave upward and downward (use the second derivative).
6. Note any points of inflection.
7. Use a table of values to plot additional points as necessary.

EXAMPLE 10.15

Sketch the graph of $f(x) = (x-1)^2(x+2)$.

SOLUTION 10.15

1. The domain is all real numbers since this is a polynomial function.

2. If $y = 0$, $(x-1)^2(x+2) = 0$, which means $x = 1$ or $x = -2$.
   If $x = 0$, $f(0) = (0-1)^2(0+2) = 2$.
   The x-intercepts are 1 and -2.
   The y-intercept is 2.

3. No asymptotes.

4. $f'(x) = (x+2)(2)(x-1) + (x-1)^2$

   Find the first derivative using the Product Rule.

   $f'(x) = (x-1)(3x+3)$ Simplify.
   $f'(x) = 0$ when $x = 1$ and $x = -1$.
   We'll find increasing and decreasing intervals and relative extrema using the first derivative:
   sign of $f'(x)$

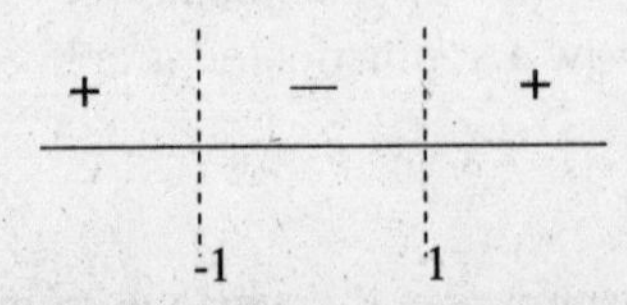

$f(x)$ is increasing on $(-\infty, -1)$ and $(1, \infty)$.
$f(x)$ is decreasing on $(-1, 1)$.
$(-1, f(-1)) = (-1, 4)$ is a relative maximum.
$(1, f(1)) = (1, 0)$ is a relative minimum.

5. It's easier to multiply out $f'(x)$ before finding $f''(x)$.
   $f'(x) = (x-1)(3x+3) = 3x^2 - 3$
   $f''(x) = 6x$
   $f''(x) = 0$ when $x = 0$
   sign of $f''(x)$

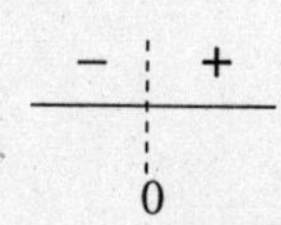

   $f(x)$ is concave downward on $(-\infty, 0)$.
   $f(x)$ is concave upward on $(0, -\infty)$.
6. $(0, f(0)) = (0, 2)$ is an inflection point.

Putting all this information together we have:

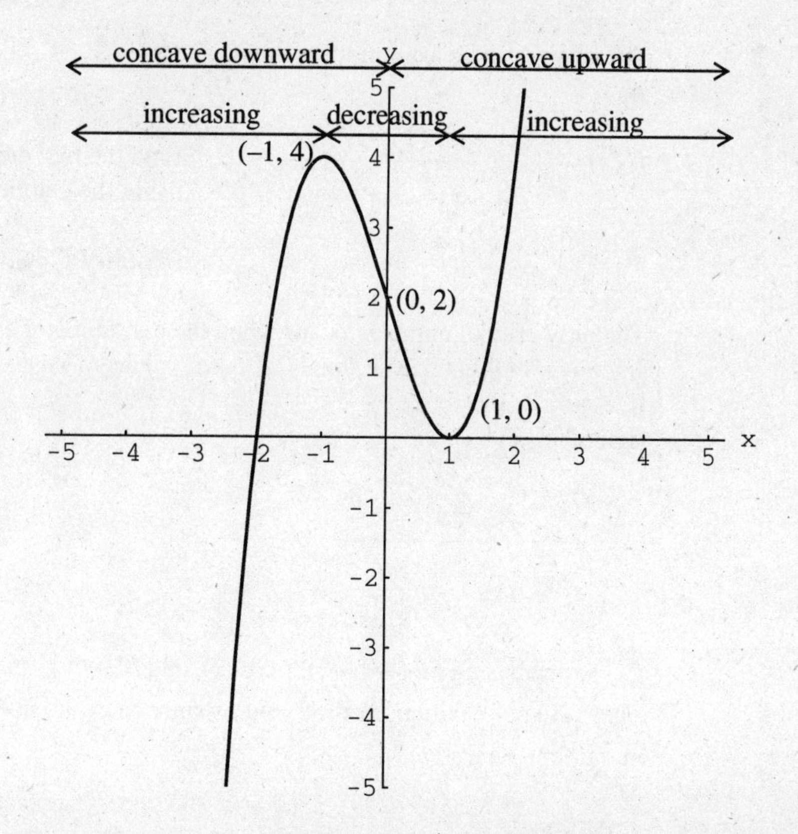

EXAMPLE 10.16

Sketch the graph of $f(x) = \dfrac{x}{x^2 - 4}$

SOLUTION 10.16

1. To find the domain, set the denominator equal to 0 to find restricted values.
   $x^2 - 4 = 0$
   $x^2 = 4$
   $x = \pm 2$
   The domain is all real numbers *except* –2 and 2.

2. If $y = 0, x = 0$.
   If $x = 0, y = 0$.
   (0, 0) is the x-intercept and the y-intercept.

3. There are vertical asymptotes at $x = 2$ and $x = -2$.

4. $f'(x) = \dfrac{(x^2 - 4)(1) - x(2x)}{(x^2 - 4)^2}$ Find the first derivative using the Quotient Rule.

   $f'(x) = \dfrac{-x^2 - 4}{(x^2 - 4)^2}$ Simplify $f'(x)$.

   The only critical numbers occur when the denominator equals 0 since $-x^2 - 4$ can never equal 0 for real values of x.
   If $x^2 - 4 = 0, x = \pm 2$
   sign of $f'(x)$

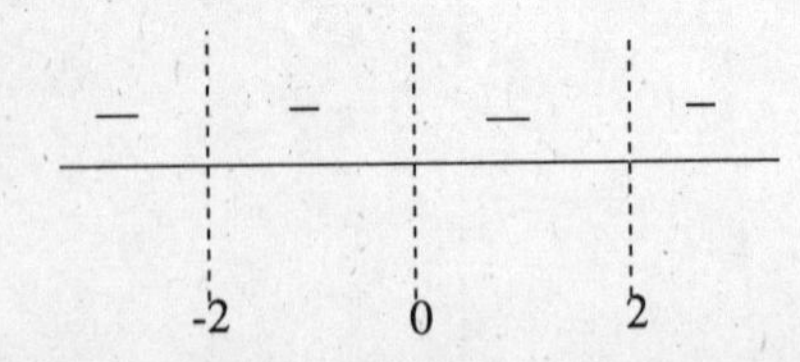

   $f(x)$ is decreasing on $(-\infty, -2)$, $(-2, -2)$, and $(2, \infty)$.
   There are no maximums or minimums since there are no sign changes in the first derivative.

5. $f''(x) = \dfrac{(x^2 - 4)^2(-2x) - (-x^2 - 4)(4x)(x^2 - 4)}{(x^2 - 4)^4}$

   Use the Quotient Rule to find the second derivative.

$$f''(x) = \frac{-2x(x^2-4)\,[(x^2-4)+(-x^2-4)(2)]}{(x^2-4)^4}$$

Factor.

$$f''(x) = \frac{-2x(x^2-4)(-x^2-12)}{(x^2-4)^4}$$

Simplify.

$$f''(x) = \frac{2x(x^2+12)}{(x^2-4)^3}$$

Reduce.

$f''(x) = 0$ when $2x(x^2+12) = 0$, which is only true when $x = 0$ since $x^2 + 12$ cannot equal 0. $f''(x)$ is undefined when $(x^2-4) = 0$, which is when $x = \pm 2$.

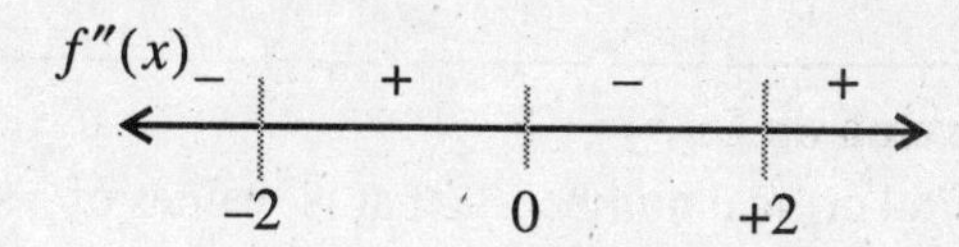

$f(x)$ is concave upward on $(-2, 0)$ and $(2, \infty)$.
$f(x)$ is concave downward on $(-\infty, -2)$ and $(0, 2)$.

6. Since the sign of $f''(x)$ changes when $x = 0$, $(0, 0)$ is an inflection point. Note that $x = -2$ and $x = 2$ cannot be x-coordinates of inflection points since they are not in the domain of $f(x)$.
Putting all this information together we have the following graph:

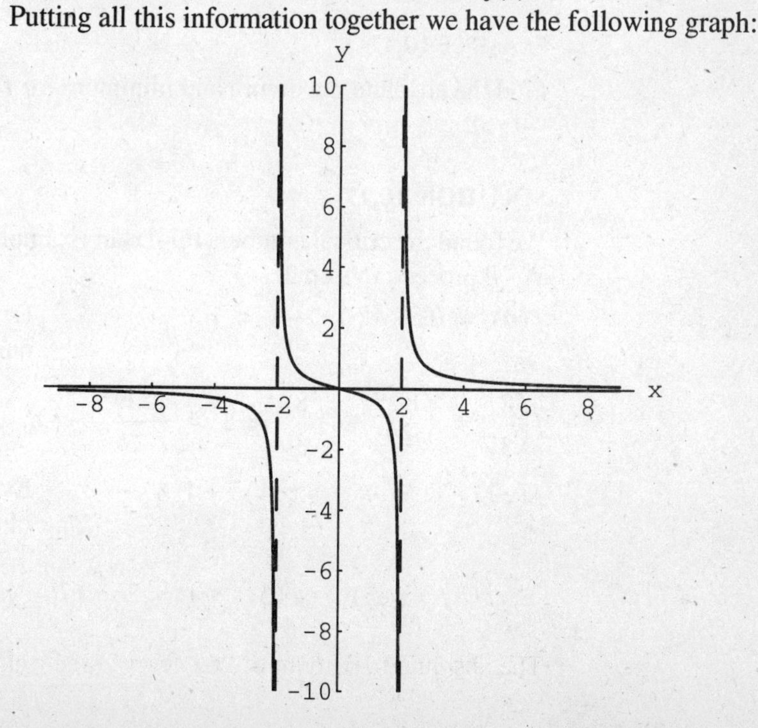

10.5 OPTIMIZATION

In this section, we will solve a type of problem for continuous functions defined on closed intervals in which either a maximum or minimum amount is needed. We can use what we know about graphs to find absolute and relative maximums and minimums, helped by a theorem that states that if a function is continuous on a closed interval, then the function will have an absolute maximum and an absolute minimum within that interval (including the endpoints of the interval.)

Extrema on a Closed Interval

If the function $f(x)$ is continuous on a closed interval $[a, b]$, we'll use the following steps to find the absolute maximum and absolute minimum.

Extrema on $[a, b]$

1. Find critical numbers—that is, values of x such that $f'(x) = 0$ or $f'(x)$ is undefined.
2. Evaluate f at each critical number.
3. Evaluate $f(a)$ and $f(b)$.
4. The largest value from steps 2 and 3 is the absolute maximum and the smallest is the absolute minimum.

EXAMPLE 10.17

Find the absolute maximum and minimum for $f(x) = x^3 - 4x^2 + 1$ on $[-1, 5]$.

SOLUTION 10.17

We found the critical numbers for $f(x)$ in Example 10.7 ($x = 0$ and $x = 8/3$). We'll proceed to step 2.

$f(0) = 0^3 - 4(0)^2 + 1 = 1$ Evaluate f at each critical number.

$$f\left(\frac{8}{3}\right) = \left(\frac{8}{3}\right)^3 - 4\left(\frac{8}{3}\right)^2 + 1 = -\frac{229}{27} \approx -8.5$$

$f(-1) = (-1)^3 - 4(-1)^2 + 1 = -4$ Evaluate f at the endpoints of the interval.

$$f(5) = (5)^3 - 4(5)^2 + 1 = 26$$

The absolute maximum of 26 occurs at the right endpoint when $x = 5$ and the absolute minimum of –229/27 occurs when $x = 8/3$.

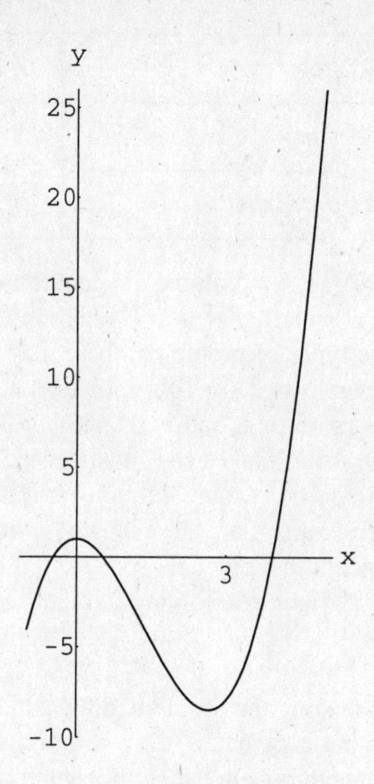

Note from the previous example that an absolute maximum or minimum may occur at either an endpoint of the closed interval *or* at a critical point.

Geometric Problems

Geometric problems usually involve maximizing an area or volume possibly subject to constraints on how the area or volume can be constructed. The use of geometric shapes usually requires that a diagram be drawn to display the given information. Also, we may need to recall the formulas for the perimeter and area of a rectangle and circle, and the formulas for the volume of a box or cylinder. These formulas follow:

Use	**Formula**
Circumference of a circle	$C = 2\pi r$
Perimeter of a rectangle	$P = 2l + 2w$
Area of a rectangle	$A = l \times w$

Use	Formula
Area of a circle	$A = \pi r^2$
Volume of a box	$V = l \times w \times h$
Volume of a cylinder	$V = \pi r^2 \times h$

where A = area, V = volume, l = length, w = width, h = height, r = radius, $\pi \approx 3.14$, P = perimeter, and C = circumference.

In these types of problems, there may be restrictions on the interval in which we will look for solutions to the problem. Also, some problems may have more than one *mathematical* solution, from which we choose the solution that makes the most sense. For example, a problem might yield two solutions, one solution representing a positive length and the other representing a negative length. Obviously, a negative length is not a useful concept in a practical sense and this solution can be discarded.

The guidelines for solving geometrical optimization problems are as follows:

(a) After reading the problem, draw a diagram and assign labels to everything that is given.
(b) Assign symbols to the significant variables in the problem.
(c) Determine the relations between the variables and try to establish an equation that defines the variable to be optimized in terms of *one* other variable.
(d) Determine the interval of interest by observing the physical and numerical *constraints* on the problem.
(e) Find the critical points for the equation formed in step c. Test answers against the interval determined in step d.

EXAMPLE 10.18

(a) A city-dweller has purchased a puppy and 400 feet of fencing. She intends to build a rectangular pen for the puppy. However, she wants the area enclosed by the fence to be as large as possible so that the puppy has room to play. What should the length and width of the pen be?
(b) A manufacturer makes steel enclosures for electrical connections. The enclosures, excluding a face plate, are to be made from a piece of metal 16 inches square. The open top is to be made by cutting a square from each corner and bending up the resulting tabs. What size square should be cut from the corner to make the volume of the elec-

trical enclosure maximum?

SOLUTION 10.18

(a)

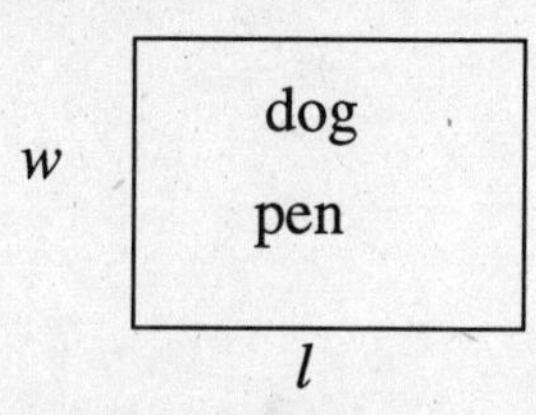

Sketch the geometric figure and label.

Area of dog pen, $A = l \cdot w$

Perimeter of dog pen, $P = 2l + 2w$

Assign variables.

length of fence = 400 feet = perimeter.

The area, A, is to be optimized, so we must define the area in terms of *one* other variable. We arbitrarily choose l.

$400 = 2l + 2w$

Set the length of fence equal to perimeter.

$200 = l + w$

Divide by 2.

$l = 200 - w$

Solve for l.

$A = l \cdot w$

$A = (200 - w) \cdot w$

Substitute for l in area formula.

$A = 200w - w^2$

Determine the physical limits of the solution interval.

l=200 ft.

w = 0 ft.

w = 200 ft.

l = 0 ft.

$0 \le w \le 200$

The width can be no narrower than 0 feet and no wider than 200 feet. (This occurs when the fencing is merely folded in half.)

$$\frac{dA}{dw} = 200 - 2w$$

Find $\frac{dA}{dw}$ and set to 0.

$0 = 200 - 2w$
$2w = 200$
$w = 100$ feet

Solve for w.

By substituting the critical point and the endpoints into the area formula, we find that $w = 100$ is where the maximum area occurs. If $w = 0$ or 200, the area enclosed is zero. All other values of w in this interval give a smaller enclosed area. Check the given solution *and* the endpoints of the interval to see which works best. Other points in the interval need *not* be checked.

$200 - w = l$
$200 - 100 = l$
$100 = l$

The length, l, is determined from the perimeter equation in part B.

$A = l \cdot w$
$A = (100) \cdot (100)$
$A = 10{,}000$ square feet

The maximum area is found by multiplying the length times the width.

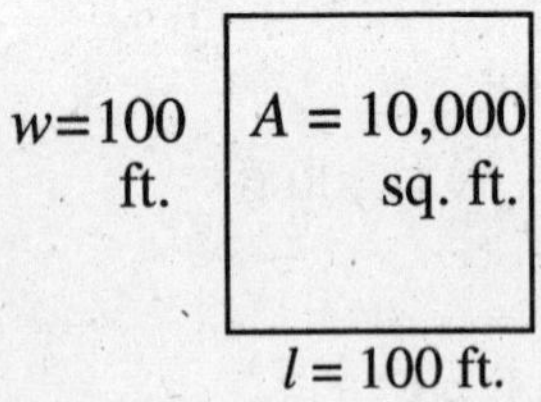

$w=100$ ft.

$A = 10{,}000$ sq. ft.

$l = 100$ ft.

The member of the rectangle family that encloses the most area for a given perimeter is a *square*. You did find that a square always gives the maximum area for a given perimeter.

(b)

Sketch the geometric figure and label.

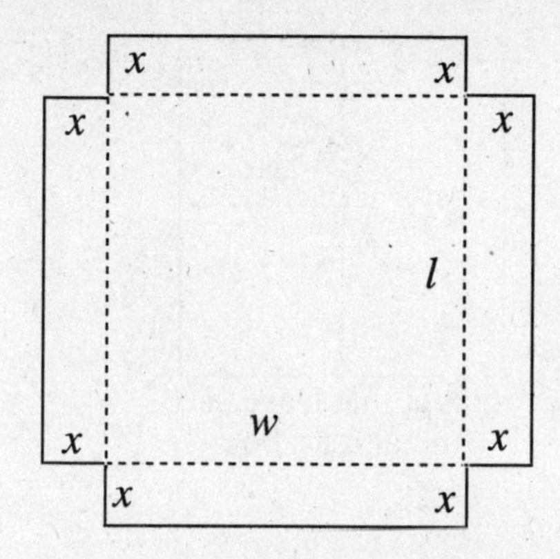

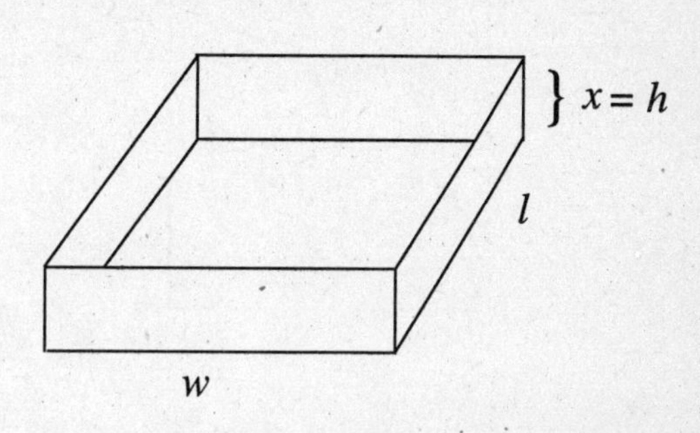

Volume of open box = $l \cdot w \cdot h$ $l = 16 - 2x = w$ $h = x$	Assign variables. The length, l, and the width, w are equal to 16 minus two times x (the length of the side of the square cutout).
$V = l \cdot w \cdot h$	The height will equal x. The volume, V, is to be optimized, so we must find the volume in terms of *one* variable.
$l = 16 - 2x$ $w = 16 - 2x$ $h = x$ $V = (16 - 2x)\ (16 - 2x)\ (x)$	The length, width, and height can be found in terms of x. Rewrite the volume in terms of x.
$V = (256 - 64x + 4x^2)\ (x)$ $V = 256x - 64x^2 + 4x^3$	
	Determine the physical limits of the solution interval for x. x can be no less than zero (no cutout) and no more than one-half of the width of the metal sheet (or 8 inches).

no cutout

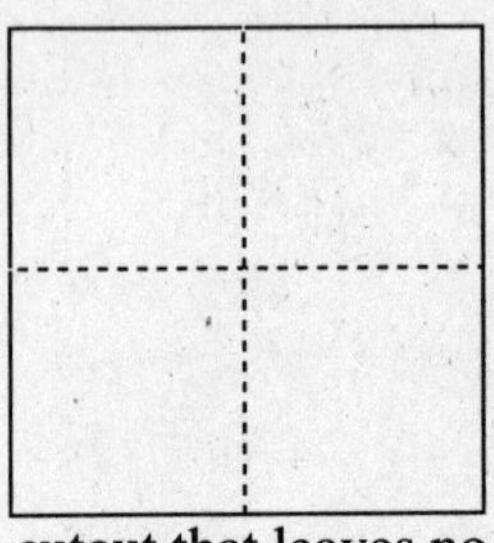

cutout that leaves no metal for a box

$$0 \le x \le 8$$

$$\frac{dV}{dx} = 256 - 128x + 12x^2$$

Find $\frac{dV}{dx}$ and set it equal to zero.

$$0 = 12x^2 - 128x + 256$$

Divide by 4.

$$0 = 3x^2 - 32x + 64$$

$$0 = (3x - 8)(x - 8)$$

Factor trinomial.

$$0 = 3x - 8 \qquad x - 8 = 0$$

$$3x = 8 \qquad x = 8$$

Solve for solution.

$$x = \frac{8}{3}$$

$$x = \frac{8}{3}$$

The value of $x = 8$ cannot be used because it leaves no metal to make a box.

$$h = x = \frac{8}{3} \text{ inches}$$

$x = 8/3$ is the optimum solution.

$$l = 16 - 2x = 16 - 2\left(\frac{8}{3}\right) = 10\frac{2}{3} \text{ inches}$$

$$w = l = 10\frac{2}{3} \text{ inches}$$

The length, height, and width can be found from relating each to x.

Note that the First or Second Derivative Test should be used to determine whether each critical value yields a maximum or minimum.

Business Applications

This class of problems is, perhaps, easier to complete since we often know the formulas to be used. The main task is to find the first derivative of the given formula, set it equal to zero, and solve. Then confirm whether the answer is a maximum or minimum using the First or Second Derivative Test. The result is the solution to the problem. The procedure for solving those problems with given formulas consists of three steps.

(1) Find the first derivative of the given formula.
(2) Set the first derivative equal to zero, and solve.
(3) Apply the First or Second Derivative Test and then compare the solution against that given by the endpoints of the interval (usually given).

EXAMPLE 10.19

(a) The profit function for the production of automobile radiators is $P(x) = 500x - x^2$ where x is the number of radiators produced each week $(0 \le x \le 500)$. How many radiators should be produced to yield the maximum profit?

(b) The concentration of a certain drug in the bloodstream at time t after ingestion is given by

$$h(t) = 6t^{2/3} - 4t \quad 0 \le t \le 3$$

where t is in hours. How many hours after ingestion is the concentration in the bloodstream maximum?

(c) Currently, the owner of a small business sells 60 wooden cabinets per week at $40 per cabinet. The owner has calculated that, for each dollar the price of the cabinet is raised, the number of cabinets sold drops by one. What price should be charged to produce the maximum revenue for the week?

SOLUTION 10.19

(a) $P(x) = 500x - x^2$ — Copy the given profit function.

$P'(x) = 500 - 2x$ — Find the first derivative of the profit function, $P(x)$.

$P'(x) = 0 = 500 - 2x$
$500 = 2x$ — Set $P'(x) = 0$ and solve.

$250 = x$ — The extreme value is 250.

Since $P''(x) = -2$, the extreme value is a maximum.

$P(0) = 500(0) - 0^2 = 0$

$P(250) = 500(250) - (250)^2$
$= 62500$

$P(500) = 500(500) - 500^2 = 0$

The endpoints are 0 and 500. The value at $P(250)$ is the largest. Therefore, the value of 250 produces the maximum profit.

(b) $h(t) = 6t^{2/3} - 4t$

Copy the given function.

$h'(t) = \frac{2}{3}(6)t^{-1/3} - 4$

$h'(t) = 4t^{-1/3} - 4$

$h'(t) = \frac{4}{t^{1/3}} - 4$

Find the first derivative of the given function.

$\frac{4}{t^{1/3}} - 4 = 0$

$\frac{4}{t^{1/3}} = 4$

$4t^{-1/3} = 4$

$t^{-1/3} = 1$

$t = 1$

Set the first derivative equal to 0 and solve.

$h(0) = 6 \cdot 0^{2/3} - 4(0)$
$= 0 - 0 = 0$

$h(1) = 6 \cdot 1^{2/3} - 4(1) = 6 - 4 = 2$

$h(3) = 6 \cdot 3^{2/3} - 4(3) = 12.52 - 12$
$= 0.52$

The given endpoints are 0 and 3. The endpoints are compared to the value of $t = 1$.

The concentration is at a maximum after 1 hour. (Confirm that $t = 1$ is a maximum using the First or Second Derivative Test.)

(c) current revenue is:
cabinets sold × price
$R = 60 \times \$40 = \$2400$

This problem does not have a given equation, but one can be found by comparing the current revenue against the change in revenue if the price is raised.

$R(x) = (60 - x)(40 + x)$ lose sale of one cabinet; add one dollar to price	The revenue function is found by multiplying the two binomials.
$R(x) = 2400 + 20x - x^2$	
$R'(x) = 20 - 2x$	Find the first derivative of the revenue function.
$20 - 2x = 0$ $20 = 2x$ $x = 10$	Set the first derivative equal to zero and solve.
$R(0) = (60)(40) = \$2400$	The limits for this function are $x = 0$, which represent no change and $x = 60$, which means no cabinets are sold.

$$R(10) = (60 - 10)(40 + 10) = (50)(50) = \$2500$$

$$R(60) = (60 - 60)(40 + 60) = (0)(100) = \$0$$

A price of $50 will produce maximum revenue.

Practice Exercises

1. Use implicit differentiation to find the derivative with respect to time.

   (a) $A = \frac{\sqrt{3}}{4}s^2$

   (b) $V = \frac{4}{3}\pi r^3$

   (c) $S = 4\pi r^2$

2. Oil spills into a lake in a circular pattern. If the radius of the circle increases at a rate of 6 inches per minute, how fast is the area of the spill increasing at the end of 1 hour? Give your answer in ft^2/min.

3. Air is being pumped into a spherical balloon at a rate of 4 cubic inches per minute. Find the rate of change of the radius when the radius is 8 inches.

4. A 10 foot ladder is leaning against the wall of a house. The base of the ladder slides away from the wall at a rate of 3 inches per second. How fast is the top of the ladder moving down the wall when the base is 5 feet from the wall?

5. Find the intervals on which the function is increasing and decreasing.

   (a) $f(x) = x^4 - 2x^2 + 1$

   (b) $f(x) = \frac{2x^2}{4x^2 - 1}$

6. Find the absolute maximum and minimumon the indicated closed interval.

   $f(x) = x^2 + 2x - 1$ on $[-2, 2]$

7. Find the relative extrema for each function.

   (a) $f(x) = \frac{1}{3}x^3 + x^2 - 3x + 2$

   (b) $f(x) = \frac{x-3}{2x+1}$

   (c) $f(x) = \frac{1}{4}x^4 - x^3 - \frac{1}{2}x^2 + 3x$

8. Determine the intervals where the function is concave upward or concave downward.

   (a) $f(x) = \frac{1}{12}x^4 - \frac{1}{2}x^3 - 5x^2$

   (b) $f(x) = \frac{4x}{x-2}$

9. Find the inflection point(s), if any, for each function.

   (a) $f(x) = \frac{1}{12}x^4 - \frac{1}{2}x^3 - 5x^2$

   (b) $f(x) = \frac{4x}{x-2}$

10. Use the Second Derivative Test to find the relative extrema for the function

    $f(x) = \frac{1}{3}x^3 - 2x^2 - 5x$

11. Use all the graphing skills developed to sketch each graph. Note the relative extrema, inflection points, asymptotes, increasing and decreasing intervals and concave upward and downward intervals.

    (a) $f(x) = \frac{1}{4}x^4 - x^3 - 2x^2$

    (b) $f(x) = \frac{x^2+1}{x^2-4}$

Answers

1. (a) $\frac{dA}{dt} = \frac{\sqrt{3}}{2}s\frac{ds}{dt}$

(b) $\frac{dV}{dt} = 4\pi r^2\frac{dr}{dt}$

(c) $\frac{dS}{dt} = 8\pi r\frac{dr}{dt}$

2. 30π ft^2/min

3. $\frac{1}{64\pi}$ in/min

4. $(-\sqrt{3})$ in/sec

5. (a) increasing (–1, 0) and (1, ∞)
decreasing (–∞, –1) and (0, 1)

(b) increasing (–∞, –1/2) and (–1/2, 0)
decreasing (0, 1/2) and (1/2, ∞)

6. absolute maximum (2, 7)
absolute minimum (–1, –2)

7. (a) relative maximum (–3, 11)
relative minimum (1, 1/3)

(b) no relative extrema

(c) relative maximum (1, 7/4)
relative minimum (–1, –9/4) and (3, –9/4)

8. (a) concave upward (–∞, –2) (5, ∞)
concave downward (–2, 5)

(b) concave upward (2, ∞)
concave downward (–∞, 2)

9. (a) $(-2, -14\frac{2}{3})$ $(5, -135\frac{5}{12})$

(b) no inflection points

10. relative maximum (–1, 8/3)

relative minimum $(5, -33\frac{1}{3})$

11. (a)

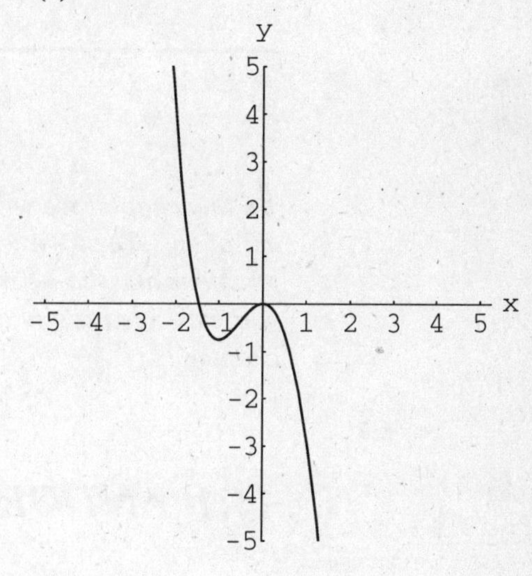

(b)

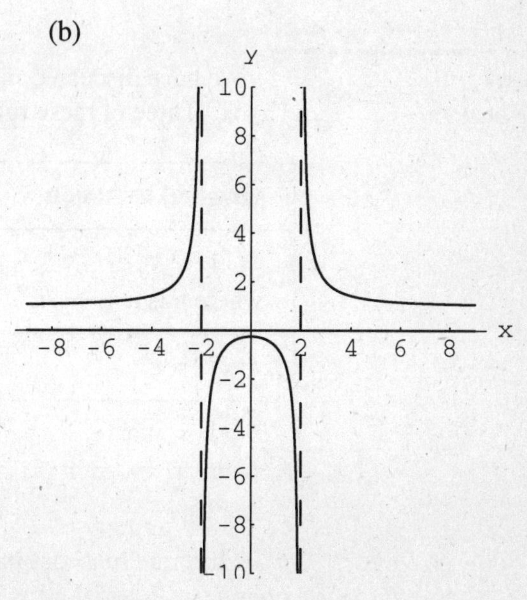

11

The Integral

In this chapter we will seek the inverse question to the process of differentiation. Then we will use this new process to find areas bounded by functions and the x-axis. Because this process will require differentiation, you may need to review parts of chapters 8 and 9 as you work through chapter 11.

11.1 ANTIDERIVATIVES

Indefinite Integrals

We have discussed many rules for derivatives in earlier sections of this book. Three of these rules are shown in Table 11.1.

Original Function	**Derivative**
$f(x) = x^n$ where n is a *numerical* exponent.	$f'(x) = nx^{n-1}$
$f(x) = e^x$	$f'(x) = e^x$
$f(x) = \ln x$	$f'(x) = \frac{1}{x}$

Table 11.1 Rules for Derivatives

The three rules outlined in Table 11.1 are the basis for our discussion of *antiderivatives*. Just as multiplication and factoring are inverse processes, so can differentiation and antidifferentiation be viewed as inverses of one another.

When we want to find the antiderivative (indefinite integral) of a function, we place that function *between* the symbols $\int$ and *dx*. The expression

$$\int f(x)\,dx$$

means "find the indefinite integral of $f(x)$." Therefore, just as we use an addition symbol, +, for addition, we will use $\int dx$ for integration.

One other comment before we state the three basic rules for integration. You may have noticed that when functions differ only by a constant, like $f(x) = x^2 - 2$ and $f(x) = x^2 + 57$, the derivative for *both* is the same, in this case, $f'(x) = 2x$. When we reverse the differentiation process, we will, in many cases, have *no idea* what the constant was *before* differentiation. For this reason, + C is added to *all indefinite integrals* because without additional information, we do not know what the constant was before the derivative was taken.

The three primary rules for indefinite integrals are listed in Table 11.2.

Rule	Original Function	Integral	Result
I	$f(x) = x^n$, $n \neq -1$	$\int x^n dx$	$F(x) = \frac{x^{n+1}}{n+1} + C$
II	$f(x) = e^x$	$\int e^x dx$	$F(x) = e^x + C$
III	$f(x) = \frac{1}{x}$	$\int x^{-1} dx$	$F(x) = \ln\|x\| + C$

Table 11.2 Rules for Indefinite Integrals

Note: The expression $F(x)$ represents the integral of some function, $f(x)$. Rule I from Table 11.2 is really the reverse of the Power Rule for derivatives. We just *add* one to the existing exponent and *divide* by the new sum. Rule II shows that the integral—like the derivative—of e^x is e^x. Rule III is for the special case when $n = -1$ for x^n. Since the derivative of $\ln x$ is $1/x$, the integral of $1/x$ is $\ln|x| + C$. The absolute value symbols are *absolutely necessary* because the logarithm of a negative quantity is undefined.

Before we begin an example of integration, we note that:

$\int k \cdot f(x)\,dx = k\int f(x)\,dx$ Constant Multiple Rule
k any real number

$\int (f(x) \pm g(x))\,dx = \int f(x)\,dx \pm \int g(x)\,dx$ Sum or Difference Rule

If $F(x) = \int f(x)\,dx$, then $F'(x) = f(x)$.

The Constant Multiple Rule says that we can factor any constant out of the integral symbol. This rule is only applicable to constants, and cannot be used to factor out x's. The Sum or Difference Rule allows us to integrate a sum or difference by separately integrating each part of the sum or difference. The third statement will be examined in section 11.3.

EXAMPLE 11.1

Find the integral (antiderivative) of the following:

(a) $f(x) = 2e^x$

(b) $f(x) = 5x^2 + 2x$

(c) $f(x) = \frac{5}{x}$

(d) $f(x) = 4x^{3/2} - 2x^{1/2}$

(e) $f(x) = 7e^x - \sqrt[3]{x}$

(f) $f(x) = 6$

(g) $f(w) = \frac{1}{w} - \frac{1}{\sqrt{w}}$

(h) $f(t) = \frac{7 - t^2}{t}$

(i) $f(x) = \frac{x - 1}{\sqrt{x}}$

SOLUTION 11.1

(a) $f(x) = 2e^x$ Copy the given function.

$\int 2e^x dx$ Rewrite as an integral.

$2\int e^x dx$ Move the constant in front of the integral symbol, $\int$.

$2e^x + C$ Use Rule II to complete integration.

Note: Only numbers (or designated constants) can be moved in front of the $\int$ symbol.

Note: Many times students are concerned about what happens to the $\int$ and the dx when the integration is completed. These symbols indicate an operation is completed and they "disappear" much like the plus sign in $2+2 = 4$.

(b) $f(x) = 5x^2 + 2x$	Copy the given function.
$\int (5x^2 + 2x)\,dx$	Rewrite as an integral.
$\int 5x^2 dx + \int 2x dx$	This integral can be separated at the plus sign.
$5\int x^2 dx + 2\int x^1 dx$	Move the constants.
$\frac{5x^{2+1}}{2+1} + \frac{2x^{1+1}}{1+1} + C$	Use Rule I to find the antiderivative (integral).
$\frac{5x^3}{3} + \frac{2x^2}{2} + C$	Simplify.
$\frac{5}{3}x^3 + x^2 + C$	Simplify the expression.
(c) $f(x) = \frac{5}{x}$	Copy the given function.
$\int \frac{5}{x} dx$	Rewrite as an integral.
$5\int \frac{1}{x} dx$	Move the constant.
$5\ln\lvert x\rvert + C$	Use Rule III to solve.
(d) $f(x) = 4x^{3/2} - 2x^{1/2}$	Copy the given function.
$\int (4x^{3/2} - 2x^{1/2})\,dx$	Rewrite as an integral.
$\int 4x^{3/2} dx - \int 2x^{1/2} dx$	Separate into two integrals.
$4\int x^{3/2} dx - 2\int x^{1/2} dx$	Move the constants.
$4\left(\frac{x^{3/2+1}}{\frac{3}{2}+1}\right) - 2\left(\frac{x^{1/2+1}}{\frac{1}{2}+1}\right) + C$	Solve using Rule I from Table 11.2.

$$4\left(\frac{x^{3/2+2/2}}{\frac{3}{2}+\frac{2}{2}}\right)-2\left(\frac{x^{1/2+2/2}}{\frac{1}{2}+\frac{2}{2}}\right)+C$$ Simplify by combining constants.

$$4\frac{x^{5/2}}{5/2}-2\frac{x^{3/2}}{3/2}+C$$

$$(4)\left(\frac{2}{5}\right)x^{5/2}-(2)\left(\frac{2}{3}\right)x^{3/2}+C$$ Invert the fractions in the denominators.

$$\frac{8}{5}x^{5/2}-\frac{4}{3}x^{3/2}+C$$ Multiply to complete solution.

(e) $f(x) = 7e^x - \sqrt[3]{x}$ Copy the given function.

$$f(x) = 7e^x - x^{1/3}$$ Convert the radical to a fractional exponent.

$$\int(7e^x - x^{1/3})\,dx$$ Rewrite as an integral.

$$\int 7e^x dx - \int x^{1/3} dx$$ Separate into two integrals.

$$7\int e^x dx - \int x^{1/3} dx$$ Move the constant.

$$7e^x - \frac{x^{1/3+1}}{\frac{1}{3}+1}+C$$ Evaluate using Rule II and Rule I from Table 11.2.

$$7e^x - \frac{x^{1/3+3/3}}{\frac{1}{3}+\frac{3}{3}}+C$$

$$7e^x - \frac{x^{4/3}}{4/3}+C$$ Add.

$$7e^x - \frac{3}{4}x^{4/3}+C$$ Invert the fraction in the denominator to obtain the solution.

(f) $f(x) = 6$ Copy the given function.

$$\int 6dx$$ Rewrite as an integral.

$\int 6x^0 dx$ — Note that $6 = 6 \cdot 1 = 6 \cdot x^0$ since $x^0 = 1$.

$6\dfrac{x^{0+1}}{0+1} + C$ — Use Rule I from Table 11.2 to evaluate the integral.

$6\dfrac{x^1}{1} + C$ — Simplify.

$6x + C$ — **Note that** $\int dx = x + C$.

(g) $f(w) = \dfrac{1}{w} - \dfrac{1}{\sqrt{w}}$ — Copy the given function.

$f(w) = \dfrac{1}{w} - \dfrac{1}{w^{1/2}}$ — Convert the radical to a fractional exponent.

$f(w) = w^{-1} - w^{-1/2}$ — Rewrite using negative exponents.

$\int (w^{-1} - w^{-1/2})\, dw$ — Rewrite as an integral.

$\int w^{-1} dw - \int w^{-1/2} dw$ — Separate into two integrals.

$\ln|w| - \dfrac{w^{-1/2+1}}{-\frac{1}{2}+1} + C$ — Use Rule III and Rule I from Table 11.2.

$\ln|w| - \dfrac{w^{-1/2+2/2}}{-\frac{1}{2}+\frac{2}{2}} + C$ — Combine the constants.

$\ln|w| - \dfrac{w^{1/2}}{\frac{1}{2}} + C$ — Invert the fraction in the denominator to obtain the final answer.

$\ln|w| - 2w^{1/2} + C$

(h) $f(t) = \dfrac{7-t^2}{t}$ — Copy the given function.

$\int \dfrac{7-t^2}{t} dt$ — Rewrite as an integral.

$\int \dfrac{7}{t} - \dfrac{t^2}{t} dt$ — Since our rules do not cover

	rational expressions, we make use of division rules.
$\int (7t^{-1} - t)\, dt$	
$\int 7t^{-1} dt - \int t dt$	Separate the integral into two parts.
$7\int t^{-1} dt - \int t dt$	Move the constant.
$7\ln\lvert t\rvert - \frac{t^2}{2} + C$	Use Rules III and I to find the antiderivative.
(i) $f(x) = \frac{x-1}{\sqrt{x}}$	Copy the function.
$f(x) = \frac{x-1}{x^{1/2}}$	Rewrite the radical as a fractional exponent.
$\int \frac{x-1}{x^{1/2}} dx$	Rewrite as an integral.
$\int \left(\frac{x}{x^{1/2}} - \frac{1}{x^{1/2}}\right) dx$	Separate into two fractions.
$\int (x^{1-1/2} - x^{-1/2})\, dx$	Divide each fraction.
$\int x^{1/2} dx - \int x^{-1/2} dx$	Separate into two integrals.
$\frac{x^{3/2}}{\frac{3}{2}} - \frac{x^{1/2}}{\frac{1}{2}} + C$	Find the antiderivative of each.
$\frac{2}{3}x^{3/2} - 2x^{1/2} + C$	Invert each fraction to obtain the answer.

Evaluating Constants of Integration

Sometimes we are able to evaluate the constant of integration, C. We can do this when we know the value of the integral at a certain point.

EXAMPLE 11.2

Find the integral and the constant of integration for the following functions.

(a) $f(x) = x + 2, \qquad F(0) = 7$

(b) $f(x) = \sqrt{x}$, $F(4) = 11$

(c) $f(x) = \frac{4}{x}$, $F(1) = 12$

(d) $f(x) = e^x + 2$, $F(0) = -7$

SOLUTION 11.2

(a) $f(x) = x + 2$, $F(0) = 7$ — Copy the given function.

$\int (x+2)\,dx$ — Rewrite as an integral.

$F(x) = \frac{x^2}{2} + 2x + C$ — Evaluate the integral.

$F(0) = \frac{0^2}{2} + 2(0) + C = 7$ — Equate the integrated function at 0 to the number 7.

$0 + C = 7$

$C = 7$ — The constant of integration is 7.

$F(x) = \frac{x^2}{2} + 2x + 7$ — The complete antiderivative is shown.

(b) $f(x) = \sqrt{x}$, $F(4) = 11$ — Copy the given function.

$f(x) = x^{1/2}$ — Convert the radical to a fractional exponent.

$\int x^{1/2} dx$ — Rewrite as an integral.

$F(x) = \frac{x^{3/2}}{\frac{3}{2}} + C = \frac{2}{3}x^{3/2} + C$ — Evaluate the integral.

$F(4) = \frac{2}{3}(4)^{3/2} + C = 11$ — Since $F(4) = 11$, substitute 4 for x.

$\frac{2}{3}(\sqrt{4})^3 + C = 11$

$\frac{2}{3}(2)^3 + C = 11$ — Solve for C.

$\frac{2}{3}(8) + C = 11$

$\frac{16}{3} + C = 11$

$C = \frac{33}{3} - \frac{16}{3}$	Use $11 = \frac{33}{3}$.
$C = \frac{17}{3}$	The constant of integration is $17/3$.
$F(x) = \frac{2}{3}x^{3/2} + \frac{17}{3}$	Determine the complete integral.
(c) $f(x) = \frac{4}{x}$, $F(1) = 12$	Copy the function.
$\int \frac{4}{x}\,dx$	Rewrite as an integral.
$4\int \frac{1}{x}\,dx$	Move the constant.
$F(x) = 4\ln\lvert x\rvert + C$	Evaluate the indefinite integral.
$F(1) = 4\ln\lvert 1\rvert + C = 12$	
$4(0) + C = 12$	$F(1) = 12$.
$C = 12$	The constant of integration is 12.
$F(x) = 4\ln\lvert x\rvert + 12$	Write the complete antiderivative.
(d) $f(x) = e^x + 2$, $F(0) = -7$	Copy the given function.
$\int (e^x + 2)\,dx$	Rewrite as an integral.
$F(x) = e^x + 2x + C$	Determine the indefinite integral.
$F(0) = e^0 + 2(0) + C = -7$	Since $F(0) = -7$, substitute 0 for x.
$1 + 0 + C = -7$	
$C = -7 - 1$	
$C = -8$	The constant of integration is -8.
$F(x) = e^x + 2x - 8$	Write the complete antiderivative.

11.2 INTEGRATION BY SUBSTITUTION

The Chain Rule for derivatives was discussed in chapter 8. The following three examples show some results of that rule.

Original Function	Derivative
$f(x) = (x^2 - 5x)^3$	$f'(x) = 3(x^2 - 5x)^2 (2x - 5)$
$f(x) = e^{x^2 - 1}$	$f'(x) = e^{x^2 - 1} (2x)$
$f(x) = \ln (x^2 - 2)$	$f'(x) = \frac{1}{x^2 - 2} (2x) = \frac{2x}{x^2 - 2}$

Table 11.3 Chain Rule Applications

Whereas the original function can be viewed as a single function, the derivative can be viewed as a product or a quotient of more than one function. In this section, we will reverse the process of differentiation when the Chain Rule was used to find the derivative.

Sometimes we can deduce the format necessary to allow a product or quotient of functions to be integrated. This can be done by letting some *part* of the integrand (function to be integrated) equal a new variable. By convention, the variable u is most often used for the substitution variable.

Table 11.2 is adapted in Table 11.4 to show how the basic three rules are modified for use with the substitution variable.

Rule	Original Function	Integral	Result
I	$f(u) = u^n$, $n \neq 1$	$\int u^n du$	$\frac{u^{n+1}}{n+1} + C$
II	$f(u) = e^u$	$\int e^u du$	$e^u + C$
III	$f(u) = \frac{1}{u}$	$\int u^{-1} du$	$\ln \lvert u \rvert + C$

Table 11.4 Integration Rules Adapted for u-Substitution

The difference between Table 11.4 and Table 11.2 is that u can stand for a polynomial, exponential, or logarithmic function of x. For example, u

can equal $3x^3 - 2x - 1$ or e^{2x} or $\ln(x^2 - 1)$ or any other similar function. The selection of the proper u-substitution is learned from experience. This selection process is not exact, and *must be done through trial and error. If you make a substitution and it doesn't work, try another substitution.*

After a demonstration of u-substitution, a list of *suggestions* will be presented to help facilitate the substitution process.

The integral of $f(x) = 2x(x^2 + 6)^5$ can be found by using Rule I in Table 11.4. We assume that this function could be the result of a Chain Rule operation for derivatives. The task at hand, then, is to find

$$\int 2x(x^2 + 6)^5 \, dx$$

by converting to an integral in terms of u. In order to establish a pattern we let $u = x^2 + 6$ (*not* $(x^2 + 6)^5$).

If

$$u = x^2 + 6$$

then

$$\frac{du}{dx} = 2x$$

In Leibniz notation, $\frac{du}{dx}$ can be treated as a fraction, thus:

$$du = 2xdx$$

If the original integral is rewritten as

$$\int \underbrace{(x^2 + 6)}_{u}{}^5 \underbrace{2xdx}_{du}$$

the substituion of u and du becomes apparent, and the integral now becomes

$$\int (u)^5 du$$

which can be easily integrated using Rule I from Table 11.4:

$$\frac{u^6}{6} + C$$

Since u was originally equated to $x^2 + 6$, the answer is

$$\int 2x(x^2 + 6)^5 \, dx = \frac{(x^2 + 6)^6}{6} + C$$

It is confusing to find a derivative within the integration process. Remember that the differentiation is used *only to find a pattern* and should

be considered as "scratchwork" done in conjunction with the integration.
Some suggestions for u-substitutions follow.

SUGGESTIONS FOR u-SUBSTITUTION

(a) If the integrand (the function to be integrated) contains a polynomial within parentheses raised to a power, let u equal the terms inside of the parentheses.

$$\text{for } 3x^2(x^3-8)^5 \qquad \text{let } u = x^3-8$$

(b) If the integrand is a radical expression, let u equal the radicand.

$$\text{for } 5\sqrt{5x-2} \qquad \text{let } u = 5x-2$$

(c) If the integrand contains e with an exponent other than x, let u equal the exponent.

$$\text{for } e^{x^2-2} \qquad \text{let } u = x^2-2$$

(d) If the integrand is a fraction, let u equal the denominator of the fraction, excluding any numerical powers of the denominator.

$$\text{for } \frac{12x-5}{6x^2-5x} \qquad \text{let } u = 6x^2-5x$$

$$\text{for } \frac{4}{(2x+1)^3} \qquad \text{let } u = 2x+1$$

Finally, it is again important to remember that only constants can be "moved across" the $\int$ symbol. A constant may be needed to complete some substitution problems. If a constant is appended to the integrand, its *reciprocal must be placed* in front of the $\int$ symbol. The net effect is as if the integrand were multiplied by 1.

EXAMPLE 11.3

Find the following integrals through substitution.

(a) $\int 3x^2(x^3+7)^6\,dx$

(b) $\int x(3x^2+1)^5\,dx$

(c) $\int x\sqrt{x^2+4}\,dx$

(d) $\int 8xe^{5x^2-1}\,dx$

(e) $\int \frac{2x-5}{x^2-5x+1}\,dx$

(f) $\int \frac{x}{\sqrt{x^2+3}}\,dx$

(g) $\int \frac{\ln x}{x}\,dx$

(h) $\int \frac{4}{x+5}\,dx$

SOLUTION 11.3

(a) $\int 3x^2(x^3+7)^6\,dx$ — Copy the integral.

$\int 3x^2 \cdot (x^3+7)^6\,dx$ — This is the product of two functions, $3x^2$ and $(x^3+7)^6$.

Scratchwork:

$u = x^3+7$ — Let $u = x^3+7$.

$\frac{du}{dx} = 3x^2$

$du = 3x^2dx$ — We need $3x^2dx$ to obtain du.

$\int \underbrace{(x^3+7)}_{u}{}^6\, \underbrace{3x^2dx}_{du}$ — Rewrite the integral to facilitate substitution.

$\int u^6du$ — The format follows Rule I of Table 11.4.

$\frac{u^7}{7}+C$ — Find the indefinite integral.

$\frac{(x^3+7)^7}{7}+C$ — Replace x^3+7 for u to get the final result.

The result can be checked through differentiation.

(b) $\int x(3x^2+1)^5\,dx$ — Copy the integral. This integral is the product of x and $(3x^2+1)^5$.

Scratchwork:

$u = 3x^2+1$ — Let $u = 3x^2+1$.

$\frac{du}{dx} = 6x$

$du = 6xdx$	We need $6xdx$ to obtain du.
$\int (3x^2+1)^5 xdx$	Rewrite the integral.
$\frac{1}{6}\int \underbrace{(3x^2+1)}_{u}{}^5 \underbrace{6xdx}_{du}$	The given integral does not contain $6xdx$, but we can produce the constant by multiplying inside the integral by 6 and outside the integral by 1/6.
$\frac{1}{6}\int u^5 du$	The integral now fits the format of Rule I, Table 11.4.
$\frac{1}{6}\left(\frac{u^6}{6}\right)+C$	Find the indefinite integral.
$\frac{u^6}{36}+C$	Multiply fractions.
$\frac{(3x^2+1)^6}{36}+C$	Substitute for u to obtain the final answer.

Check through differentiation.

(c) $\int x\sqrt{x^2+4}dx$	Copy the given integral.
$\int x(x^2+4)^{1/2} dx$	Rewrite the radical as a fractional exponent.
Scratchwork:	The integrand can be viewed as the product of two functions.
$u = x^2+4$	Let $u = x^2+4$.
$\frac{du}{dx} = 2x$	We need $2xdx$ to obtain a du.
$du = 2xdx$	
$\int (x^2+4)^{1/2} xdx$	Rewrite the integral.
$\frac{1}{2}\int \underbrace{(x^2+4)}_{u}{}^{1/2} \underbrace{2xdx}_{du}$	A 2 can be included in the integrand if a $1/2$ is placed before the $\int$ symbol.
$\frac{1}{2}\int u^{1/2} du$	The format follows Rule I of

	Table 11.4.
$\frac{1}{2}\left(\frac{u^{3/2}}{3/2}\right)+C$	Find the indefinite integral.
$\frac{1}{2}\cdot\frac{2}{3}\cdot(u^{3/2})+C$	Simplify the form.
$\frac{1}{3}u^{3/2}+C$	
$\frac{1}{3}(x^2+4)^{3/2}+C$	Replace u with x^2+4.

Check the result through differentiation.

(d) $\int 8xe^{5x^2-1}dx$	Copy the given integral.
$\int 8xe^{5x^2-1}dx$	The integrand is the product of $8x$ and e^{5x^2-1}.

Scratchwork:

$u = 5x^2 - 1$	Let u equal the exponent of e.
$\frac{du}{dx} = 10x$	
$du = 10xdx$	
$\int e^{5x^2-1}8xdx$	Rewrite the order of the integrand.
$8\int e^{5x^2-1}xdx$	Our scratch work calls for $10x$, not $8x$. Move the 8 before the $\int$ symbol.
$8\cdot\frac{1}{10}\int \overbrace{e^{5x^2-1}}^{u}\underbrace{10xdx}_{du}$	The constant, 10, is placed before the xdx and $1/10$ is placed outside of the integral.
$\frac{8}{10}\int e^u du$	The format follows Rule II of Table 11.4.
$\frac{4}{5}e^u+C$	The antiderivative is determined ($8/10$ was reduced to $4/5$).

$\frac{4}{5}e^{5x^2-1}+C$	Let $u = 5x^2 - 1$ to obtain the final result.
(e) $\int \frac{2x-5}{x^2-5x+1}dx$	Copy the given integral.
Scratchwork:	
$u = x^2 - 5x + 1$ $\frac{du}{dx} = 2x - 5$	The suggested technique is to let u equal the denominator, $x^2 - 5x + 1$.
$du = (2x-5)\,dx$	We need $(2x-5)dx$ to obtain du.
$\int \underbrace{\frac{1}{x^2-5x+1}}_{u} \underbrace{(2x-5)\,dx}_{du}$	Rewrite the inegral.
$\int \frac{1}{u}du$	The format follows Rule III of Table 11.4.
$\ln\lvert u\rvert + C$	The antiderivative is determined.
$\ln\lvert x^2 - 5x + 1\rvert + C$	Let $u = x^2 - 5x + 1$ to obtain the final result.

Check the result through differentiation.

(f) $\int \frac{x}{\sqrt{x^2+3}}dx$	Copy the integral.
Scratchwork:	
$u = x^2 + 3$ $\frac{du}{dx} = 2x$	The suggested technique is to let the u equal the contents of the radical in the denominator.
$du = 2xdx$	We need $2xdx$ to obtain du.
$\int \frac{x}{(x^2+3)^{1/2}}dx$	Rewrite the radical sign as a fractional exponent.
$\int (x^2+3)^{-1/2}\,xdx$	Convert the exponent in the denominator to a negative exponent in the numerator.
$\frac{1}{2}\int (x^2+3)^{-1/2}\,2xdx$	We need $2xdx$, so place a 2

	in the integrand and place $1/2$ before $\int$.
$\frac{1}{2}\int \underbrace{(x^2+3)^{-1/2}}_{u} \underbrace{2x\,dx}_{du}$	Rewrite the integral.
$\frac{1}{2}\int u^{-1/2}\,du$	Use Rule I of Table 11.4.
$\frac{1}{2}\left(\frac{u^{1/2}}{1/2}\right)+C$	Find the integral.
$\frac{1}{2}\cdot\frac{2}{1}(x^2+3)^{1/2}+C$	Replace u with x^2+3 to obtain the final result.
$(x^2+3)^{1/2}+C$	
(g) $\int \frac{\ln x}{x}\,dx$	Copy the integral.
Scratchwork: (Trial 1) $u = x$	Normally, we might let u equal the denominator, x.
$\frac{du}{dx} = 1$	
$du = dx$	Note that the value of du differs from $\ln x\,dx$ by *more than just a constant*. Therefore, this choice of u-substitution will not work.
Scratchwork: (Trial 2) $u = \ln x$	
$\frac{du}{dx} = \frac{1}{x}$	
$du = \frac{1}{x}dx$	Our second trial has $du = \frac{1}{x}dx$.
$\int \underbrace{(\ln x)}_{u} \underbrace{\frac{1}{x}dx}_{du}$	Rewrite the integral.
$\int u\,du$	Use Rule I, Table 11.4.
$\frac{u^2}{2}+C$	Find the antiderivative.

$$\frac{(\ln x)^2}{2} + C$$ $u = \ln x$

Check the final result through differentiation.

Our example of a wrong choice for the u-substitution is typical of this trial-and-error process.

(h) $\int \frac{4}{x+5} dx$ Copy the integral.

$4\int \frac{1}{x+5} dx$ Move the constant.

Scratchwork:

$u = x+5$ Let $u = x + 5$.

$$\frac{du}{dx} = 1$$

$du = dx$ Our du value must equal dx.

$$4\int \underbrace{\frac{1}{x+5}}_{u} \underbrace{dx}_{du}$$ Rewrite the integral.

$4\int \frac{1}{u} du$ The integrand follows the format of Rule III in Table 11.4.

$4\ln|u| + C$

$4\ln|x+5| + C$ Replace u with $x + 5$.

Check the result through differentiation.

11.3 THE DEFINITE INTEGRAL AND THE FUNDAMENTAL THEOREM OF CALCULUS

Perhaps, when you have been confronted with the task of finding the area of an irregularly shaped room or plot of land, you decided to break the area into little rectangles, triangles or circles. The total area could then be found by adding all the pieces together.

Consider the area bounded by the curve, $f(x)$, vertical lines $x = a$, and $x = b$, and the x-axis, as in Figure 11.1.

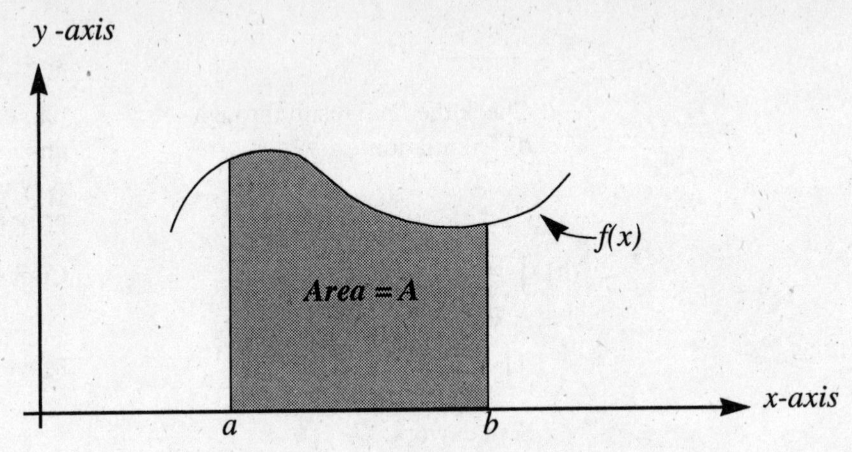

Figure 11.1 Area Under a Curve

The major problem in finding the area, A, is that the function, $f(x)$, forms a very irregular upper border. We can approximate the area by dividing it into several rectangles of *equal* width. We refer to the width as Δx (for change in x).

Figure 11.2 shows how the area might by divided.

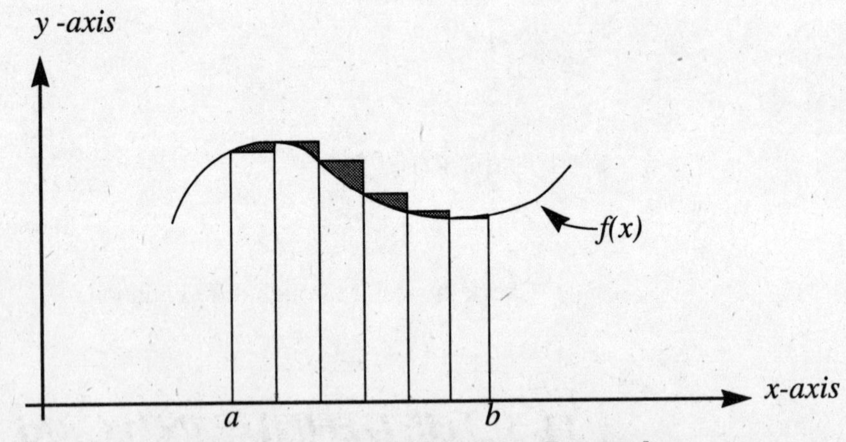

Figure 11.2 Area Divided into Rectangles

There are six rectangles that nearly sum to the area, A, except for the error indicated by the darkly shaded areas. We have already stated that each rectangle is the *same width*. The height of each rectangle is determined by where we are on the curve, $f(x)$. Note that not all rectangles are the same height but all have one top corner touching the curve. Figure 11.3 shows the closeup of one rectangle.

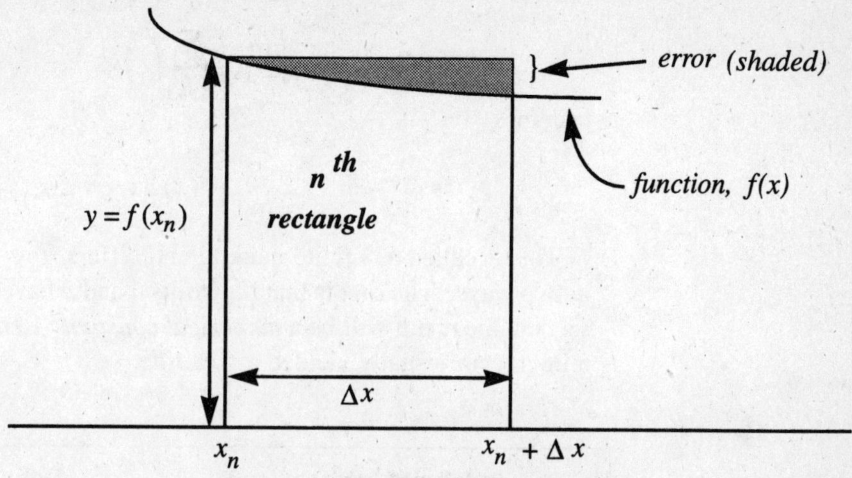

Figure 11.3 Closeup of Rectangle

The area of a rectangle is known to be length times width, so for our rectangles of length $f(x)$ and width Δx, we have an area, $A_h = f(x_h) \cdot \Delta x$. If we use sigma notation to indicate the sum of all N rectangles A_k for $k = 1$ to N, the total area is

$$\sum_{k=1}^{N} f(x_k)\,\Delta x.$$

where $f(x_k)$ is subscript notation and k labels the x-value used to compute the height we are talking about.

It seems reasonable to suppose our error will diminish if we increase the number of rectangles, and make them very narrow at the same time. See Figure 11.4.

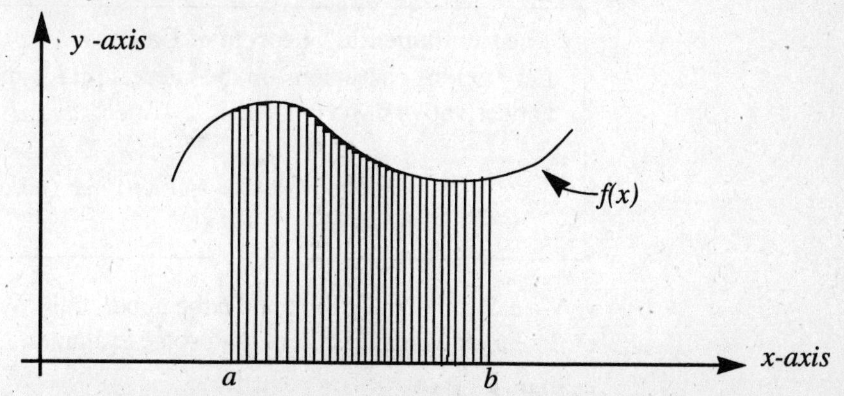

Figure 11.4 Approximating Area with Many Rectangles

Now if we allow Δx to approach 0 (dx in Leibniz notation), that is, take the limit as Δx approaches 0, then

$$\lim_{N \to \infty} \sum_{k=1}^{N} f(x_k)\,\Delta x$$

becomes

$$\int_a^b f(x)\,dx$$

This is called a definite integral and differs from the indefinite integral in two ways. The first is that the limits a and b have been introduced, and, second, the result will be a *numerical constant*. Remember that the indefinite integral usually yielded a *function*.

Indefinite Integral	**Definite Integral**
No Limits of integration	Limits of integration
Result is a *function*	Result is a *constant*
$+C$ is needed	No $+C$ needed

Table 11.5 Differences Between an Indefinite Integral and a Definite Integral

The definition just given for a definite integral and the definition of antiderivative give rise to The Fundamental Theorem of Calculus.

The Fundamental Theorem of Calculus

Let $f(x)$ be continuous on the interval $[a, b]$, and let $F(x)$ be the antiderivative of $f(x)$. Then,

$$\int_a^b f(x)\,dx = F(b) - F(a)$$

We call a the lower limit and b the upper limit. We demonstrate the use of the Fundamental Theorem with some examples.

EXAMPLE 11.4

Evaluate the following definite integrals.

(a) $\int_2^4 2x\,dx$

(b) $\int_{-1}^{3} (x^2 + 5x)\, dx$

(c) $\int_{3}^{5} \frac{5}{x} dx$

(d) $\int_{0}^{2} 2e^x dx$

SOLUTION 11.4

(a) $\int_{2}^{4} 2xdx$ — Copy the integral.

$2\int_{2}^{4} xdx$ — Move the constant.

$\left.\frac{2x^2}{2}\right|_2^4$ — Find the antiderivative and retain the limits by using a vertical bar on the right side. The $\int$ and dx are no longer needed.

$\left.\frac{2x^2}{2}\right|_2^4$ — Cancel the 2's.

$\left.x^2\right|_2^4 = (4)^2 - (2)^2 = F(4) - F(2)$ — Substitute the upper limit 4 for x, then substitute the lower limit 2 for x, and subtract.

$= 16 - 4 = 12$ — The value of the definite integral is 12.

Note: The antiderivative is evaluated by computing the *upper limit* (4, here) *minus the lower limit* (2, here).

Note: Many students ask, "What happened to the + C?" If we look at the previous example, we note the $F(x) = x^2 + C$ (if we had used the indefinite integral). Now,

$$F(4) = 4^2 + C = 16 + C \qquad \text{and} \qquad F(2) = 2^2 + C = 4 + C$$

$$F(4) - F(2) = (16 + C) - (4 + C) = 16 + C - 4 - C$$

$$= 16 - 4 + 0$$ The C's cancel out!

$$= 12$$

The C's will *always cancel*, so we don't bother with them.

(b) $\int_{-1}^{3} (x^2 + 5x)\, dx$ — Copy the integral.

$\left(\frac{x^3}{3} + \frac{5x^2}{2}\right)\Big|_{-1}^{3}$ — Find the antiderivative.

$F(3) \quad - \quad F(-1)$

$\left[\left(\frac{3^3}{3} + \frac{5(3)^2}{2}\right) - \left(\frac{(-1)^3}{3} + \frac{5(-1)^2}{2}\right)\right]$

Evaluate the function at both limits and subtract.

$\left(\frac{27}{3} + \frac{5 \cdot 9}{2}\right) - \left(\frac{-1}{3} + \frac{5 \cdot 1}{2}\right)$ — Use the order of operations to simplify.

$\left(9 + \frac{45}{2}\right) - \left(-\frac{1}{3} + \frac{5}{2}\right)$

$\left(\frac{18}{2} + \frac{45}{2}\right) - \left(-\frac{2}{6} + \frac{15}{6}\right)$ — Find least common denominators.

$\frac{63}{2} - \frac{13}{6}$

$\frac{189}{6} - \frac{13}{6} = \frac{176}{6}$ — The value of the definite integral is 176/6.

(c) $\int_{3}^{5} \frac{5}{x}\, dx$ — Copy the integral.

$5\int_{3}^{5} \frac{1}{x}\, dx$ — Move the constant.

$5\ln\lvert x\rvert \Big\|_3^5$	The antiderivative is the natural logarithm.
$F(5) - F(3)$ $5\ln\lvert 5\rvert - 5\ln\lvert 3\rvert$	Evaluate at the two limits.
$5(\ln 5 - \ln 3)$	The 5 is factored out.
$5\left(\ln\frac{5}{3}\right)$	Apply the rules of logarithms to rewrite the expression in parentheses as an expression with one log.
$5\left(\ln\frac{5}{3}\right)$ $5(0.511) \cong 2.555$	We can leave this as the final answer, or we can approximate using a calculator.
(d) $\int_0^2 2e^x dx$	Copy the integral.
$2\int_0^2 e^x dx$	Move the constant.
$2e^x\Big\|_0^2$	Find the antiderivative.
$F(2) - F(0)$ $2e^2 - 2e^0$	Evaluate the integral at the limits and subtract.
$2e^2 - 2(1)$	
$2e^2 - 2$	The answer can be left in this form or approximated using decimals.

$\approx 2(7.389) - 2$

$= 15.778 - 2 = 13.778$

11.4 APPLICATIONS OF INTEGRATION

The applications discussed in this section fall into two areas (of many possible):

(a) Area under the curve or total consumption
(b) Consumer surplus

Area and Total Consumption

The evaluation of the definite integral is the procedure needed for solving area and total consumption problems. The evaluation of the area or of the total consumption is based on the region bounded by the limits of integration, the x-axis, and the function under consideration.

EXAMPLE 11.5

(a) Find the area bounded above by the function $f(x) = 2x$, below by the x-axis, and lying between $x = 0$ and $x = 4$.

(b) Find the total area between $f(x) = x^2 - 4$ and the x-axis from $x = 0$ to $x = 3$.

(c) The rate of consumption of oil by a small country is

$$C'(t) = 1.5e^{0.10t}.$$

The consumption is in millions of barrels and t is in years. How much oil does the country consume in 10 years?

SOLUTION 11.5

(a)

A drawing of the region in question is very useful.

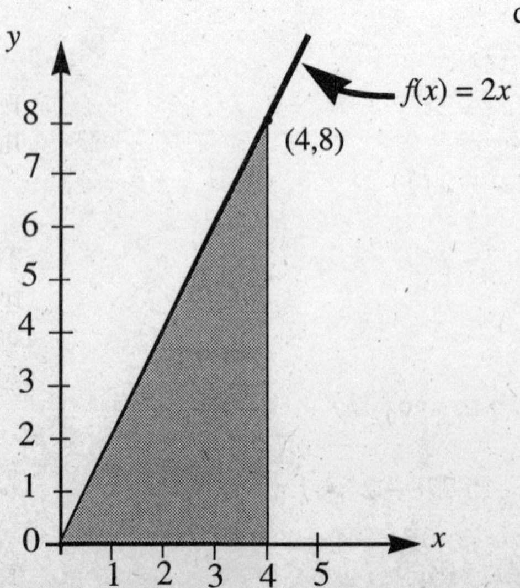

$$A = \int_0^4 2x\,dx$$

We are summing an infinite number of rectangular areas from $x = 0$ to $x = 4$. Thus, an integral is needed. (A = area)

$$A = x^2\Big|_0^4$$

The integral is determined and evaluated at the limits of integration.

$$A = 4^2 - 0^2 = 16 - 0$$

$$A = 16$$

The area of this region is 16.

Note: The region has a triangular shape with base (along the x-axis) of 4 units and height of 8 units. The area of a triangle can be determined by the formula $A = (1/2)\,bh$. Since b = base = 4 and h = height = 8 for this triangle, Area = (1/2)(4)(8) = (1/2)(32) = 16. The area found through the triangle area formula is the *same* as that found through integration.

(b)

A drawing of the region in question is *very* necessary.

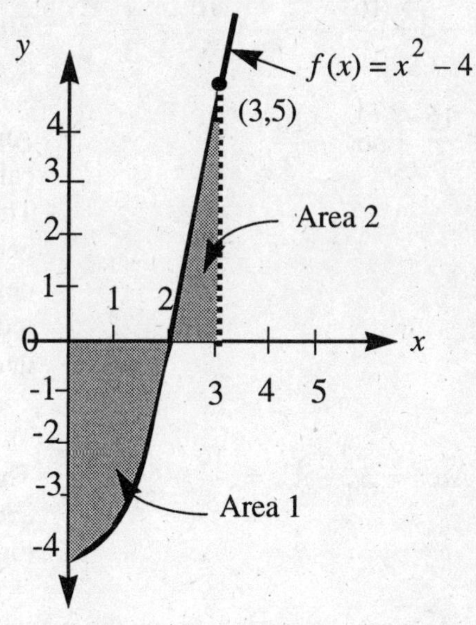

$$A_1 = \int_0^2 (x^2 - 4)\,dx$$

$$A_2 = \int_2^3 (x^2 - 4)\,dx$$

There are actually two regions in this problem, the region below the x-axis and the one above. We evaluate the *same* integrand between *two sets* of limits.

$$A_1 = \frac{x^3}{3} - 4x \Big|_0^2$$

$$= \left(\frac{2^3}{3} - 4(2)\right) - \left(\frac{0}{3} - 4(0)\right)$$

$$= \left(\frac{8}{3} - 8\right) - (0) = \frac{8}{3} - \frac{24}{3} = -\frac{16}{3}$$

The area of region one, A_1, is *negative*.

$$A_2 = \frac{x^3}{3} - 4x \Big|_2^3$$

$$= \left(\frac{3^3}{3} - 4(3)\right) - \left(\frac{2^3}{3} - 4(2)\right)$$

$$= \left(\frac{27}{3} - 12\right) - \left(-\frac{16}{3}\right) = (9 - 12) + \frac{16}{3}$$

$$= -3 + \frac{16}{3} = -\frac{9}{3} + \frac{16}{3} = \frac{7}{3}$$

The area of region two, A_2, is *positive*.

$$A_1 = \frac{16}{3} \left(\text{not } -\frac{16}{3}\right)$$

An area cannot, in a practical sense, be negative. Therefore, we take $|A_1|$ to get a *positive value*. The negative sign is simply an indication that the area in question lies below the x-axis.

$$\text{Total Area} = A_1 + A_2 = \frac{16}{3} + \frac{7}{3} = \frac{23}{3}$$

The total area is found by adding the two regions together.

Note: Any area occurring *below* the x-axis will be *negative* if $F(b)$ is less than $F(a)$. All we need to do is take the area as $|F(b) - F(a)|$.
Note: In this problem we split the area in question into two areas, one below the x-axis and one above the x-axis. The graph allowed us to see that the function, $x^2 - 4$, crossed the x-axis at 2. The number, 2, then became the upper limit for A_1 and the lower limit for A_2. If we had integrated from 0 to 3 directly we would have gotten an erroneous value of 3 (actually –3) for the total area. In effect, the lower area

would have subtracted the upper area out. That is why it *is very important to draw a graph*. We would not have known to split the problem into two integrals or to add areas if the graph were not present.

(c) $C'(t) = 1.5e^{0.10t}$

The consumption rate is given.

The graph of the function is below. The *area*, *A*, can be interpreted as the *total consumption* for 10 years.

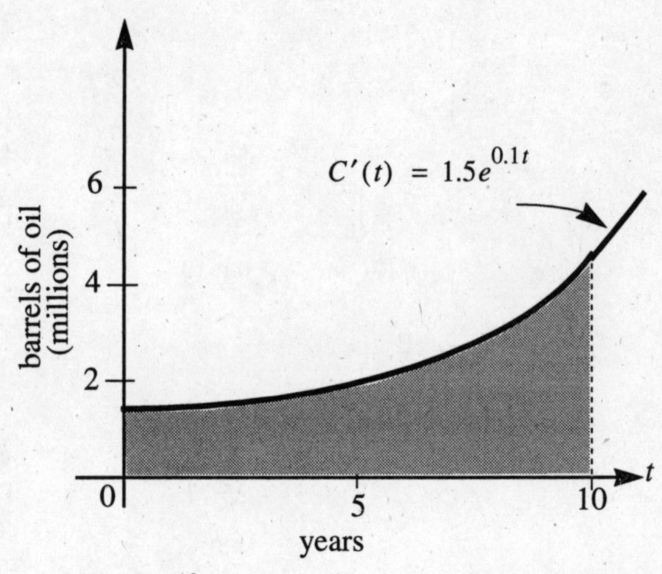

$$\text{consumption} = \int_0^{10} 1.5e^{0.10t}\,dt$$

The integral is established.

let $u = 0.10t$

$$\frac{du}{dt} = 0.10$$

$$du = 0.10dt$$

We let u equal the exponent. du must equal $0.01t$.

$$1.5\int_0^{10} e^{0.10t}\,dt$$

We need 0.10 in front of *dt*. Place 0.10 next to *dt* and put 10 outside the $\int$ symbol. $(0.10 \times 10 = 1)$

$$1.5\,(10)\int_0^{10} e^{\overbrace{0.10t}^{u}}\,\underbrace{0.10dt}_{du}$$

$$15\int e^{u}du$$

The pattern follows Rule II of Table 11.4. (The limits are temporarily not shown.)

$$= 15e^{u} + C$$

$$= 15e^{0.10t} + C$$

The integral is found and $u = 0.10t$ is replaced.

$$= 15e^{0.10t}\Big|_0^{10}$$

The limits are returned.

$$= 15e^{0.10\,(10)} - 15e^{0.10\,(0)}$$

The definite integral is evaluated.

$$= 15e^{1} - 15e^{0}$$

$$= 15e - 15\,(1) \;=\; 15e - 15$$

$$\cong 15\,(2.718) - 15 \cong 25.77$$

The result is the total oil consumption for 10 years (in millions of barrels).

Total consumption = 25.77 million barrels of oil.

Consumer Surplus

We are delighted to find a bargain. When we pay less than we expect to pay for an item, the difference is called a *consumer surplus*.

The formula for consumer surplus is:

Consumer Surplus Formula

$$\text{Consumer Surplus} = \int_0^{x_0} (p(x) - p_0)\, dx$$

where:

x_0 is the number of items produced at the equilibrium price
$p(x)$ is the demand function
p_0 is the equilibrium price

EXAMPLE 11.6

What is the consumer surplus for the demand function

$$p(x) = 200 - 10x - x^2$$

if the equilibrium price occurs at the production of 5 units per day?

SOLUTION 11.6

$$p(x) = 200 - 10x - x^2$$

Copy the demand function.

$$p(5) = 200 - 10(5) - 5^2$$
$$= 200 - 50 - 25$$
$$= 125 = p_0$$

The equilibrium price is found by substituting $x = 5$ into the demand function.

$$\text{Consumer Surplus} = \int_0^{x_0} (p(x) - p_0)\, dx$$

Use the Consumer Surplus formula to set up the integral.

$$= \int_0^5 (200 - 10x - x^2 - 125)\, dx$$

$$= \int_0^5 (75 - 10x - x^2)\, dx$$

$$= 75x - \frac{10x^2}{2} - \frac{x^3}{3}\Bigg|_0^5$$

The integral is determined.

$$= 75x - 5x^2 - \frac{x^3}{3}\Big|_0^5$$

$$= \left(75(5) - 5(5)^2 - \frac{(5)^3}{3}\right) - \left(75(0) - 5(0)^2 - \frac{0^3}{3}\right)$$

The limits are substituted.

$$= \left(375 - 125 - \frac{125}{3}\right) - 0$$

The substitution is simplified.

$$= (250 - 41.67)$$

$= \$208.33 =$ consumer surplus

The consumer surplus is determined.

Practice Exercises

1. Find the integral.

(a) $f(x) = x^2 + 7x - 2$

(b) $f(x) = 4x^{1/2} + 5x^{3/2}$

(c) $f(r) = \dfrac{r^2 + 8r}{r}$

(d) $f(x) = \dfrac{x^5}{\sqrt[3]{x}}$

2. Find the integral and the constant of integration for the following functions.

(a) $f(x) = \dfrac{6}{x} \quad F(1) = -7$

(b) $f(x) = x^2 - 3x + 1 \quad F(0) = 9$

3. Find the following integrals through substitution.

(a) $\int 2x(x^2 + 1)\,dx$

(b) $\int 3x^2\sqrt{x^3 + 2}dx$

(c) $\int \dfrac{3x^2 + 7}{x^3 + 7x + 2}dx$

(d) $\int \dfrac{6}{x + 3}dx$

(e) $\int 7xe^{x^2}dx$

4. Evaluate the following definite integrals.

(a) $\int_1^5 3x\,dx$

(b) $\int_{-2}^{2} (x^2 - x)\,dx$

(c) $\int_1^3 5e^x\,dx$

5. (a) Find the area bounded by $4 - x^2$ and the x-axis.

(b) Find the area bounded by $x = 3$, $x = 5$, $y = x^3$ and the x-axis.

Answers

1. (a) $F(x) = \frac{x^3}{3} + \frac{7x^2}{2} - 2x + C$

   (b) $F(x) = \frac{8x^{3/2}}{3} + 2x^{5/2} + C$

   (c) $F(r) = \frac{r^2}{2} + 8r + C$

   (d) $F(x) = \frac{3x^{17/3}}{17} + C$

2. (a) $F(x) = 6\ln|x| - 7$

   (b) $F(x) = \frac{x^3}{3} - \frac{3x^2}{2} + x + 9$

3. (a) $\frac{(x^2+1)^2}{2} + C$

   (b) $\frac{2\sqrt{(x^3+2)^3}}{3} + C$

   (c) $\ln\left|x^3 + 7x + 2\right| + C$

   (d) $6\ln|x+3| + C$

   (e) $\frac{7}{2}e^{x^2} + C$

4. (a) 36

   (b) 16/3

   (c) $5(e^3 - e)$

5. (a) 32/3

   (b) 136

12

Multivariable Calculus

It is important to realize that although our work to this point has involved functions of one variable, the topics of differentiation and integration can be extended to functions of several variables. We will use this final chapter to examine such multivariable functions, broadly applying some of the topics we have seen in previous chapters.

12.1 FUNCTIONS OF SEVERAL VARIABLES

Recall that a function of one variable takes the form $f(x) =$ ________ where the blank contains an expression involving x. We extend this notation to more than one variable using $f(x, y) =$ ________ where the blank now contains an expression involving x and y. If $f(x) = x^2 + 2$, $f(3)$ means replace x with 3. If $f(x, y) = x^2 + 2xy + 5$, $f(1, 3)$ means replace x with 1 and y with 3:

$$f(1, 3) = (1)^2 + 2(1)(3) + 5 = 12$$

EXAMPLE 12.1

Evaluate the given function at the indicated value.

(a) $f(x, y) = x + 2\sqrt{y}$ at $(3, 9)$

(b) $f(x, y) = x^2 - 2xy + 3y^2$ at $(-1, 2)$

(c) $f(x, y, z) = xy + 4xz - 6yz$ at $(1, 2, 3)$

SOLUTION 12.1

(a) $f(x, y) = x + 2\sqrt{y}$ Copy the function.

$$f(3, 9) = 3 + 2\sqrt{9}$$ Substitute $x = 3$, $y = 9$.

$$= 3 + 2(3)$$ Simplify.

$$= 9$$

(b) $f(x, y) = x^2 - 2xy + 3y^2$ Copy the function.

$$f(-1, 2) = (-1)^2 - 2(-1)(2) + 3(2)^2$$ Substitute $x = -1$, $y = 2$.

$$= 1 + 4 + 3(4)$$ Simplify.

$$= 17$$

(c) $f(x, y, z) = xy + 4xz - 6yz$ Copy the function.

$$f(1, 2, 3) = (1)(2) + 4(1)(3) - 6(2)(3)$$ Substitute $x = 1$, $y = 2$, and $z = 3$.

$$= 2 + 12 - 36$$ Simplify.

$$= -22$$

EXAMPLE 12.2

A company produces two types of calculators, nonprogrammable and programable. The cost function in dollars to produce n nonprogrammable calculators and p programmable calculators is given by

$$C(n, p) = 10000 + 50n + 200p$$

Find the cost to produce 1000 nonprogrammable calculators and 500 programmable calculators.

SOLUTION 12.2

We are asked to find $C(1000, 500)$.

$$C(1000, 500) = 10000 + 50(1000) + 200(500)$$
$$= 10000 + 50000 + 100000$$
$$= \$160{,}000$$

Graphing $f(x, y)$

The graph of $z = f(x, y)$ is the set of ordered triples (x_0, y_0, z_0) that satisfy the given equation. Graphing three dimensions on a two dimensional object such as a piece of paper is difficult. We use three axes, labeled x, y, and z, and draw them to make the x axis "appear" to come out of the paper:

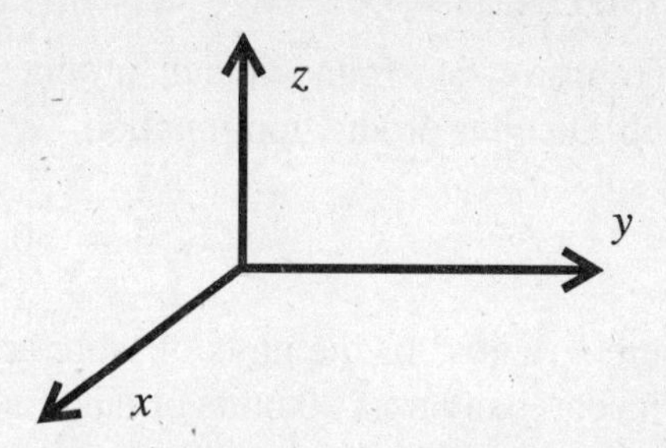

Points are graphed by moving the indicated amount parallel to the axes. For example, (1, 2, 5) would be drawn as

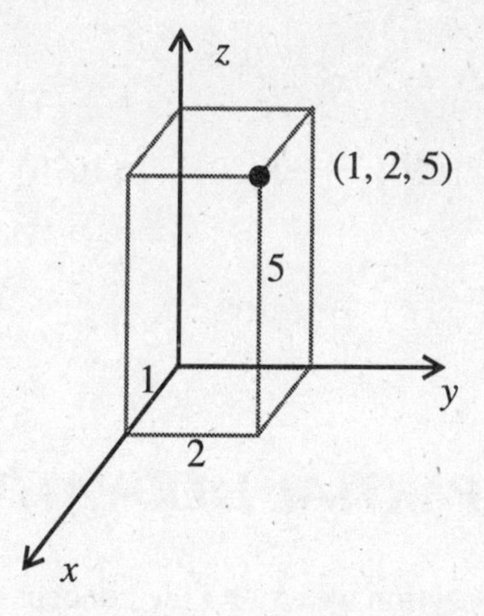

Note that (1, 2, 5) is the vertex of a box with width 1 (x-value), length 2 (y-value), and height 5 (z-value). Most courses at this level do not emphasize this type of graphing due to the difficulty involved.

Cobb-Douglas Production Function

In general, a production function represents the relationship between production inputs such as materials and labor, and the quantity produced.

In particular, the Cobb-Douglas production function relates labor and capital, and can be expressed in the form

$$f(x, y) = ax^{\lambda}y^{1-\lambda}$$

where $f(x, y)$ is the number of units produced from x units of labor and y units of capital, and a, λ, and $1 - \lambda$ are constants.

EXAMPLE 12.3

A company has found it can produce calculators as described by the Cobb-Douglas production function

$$f(x, y) = 30x^{0.25}y^{0.75}$$

where x represents the units of labor and y represents the units of capital. If the company uses 20 units of labor and 1620 units of capital, how many calculators will be produced?

SOLUTION 12.3

We are asked to find $f(20, 1620)$.

$$f(x, y) = 30x^{0.25}y^{0.75}$$ Copy the given equation.

$$f(20, 1620) = 30(20)^{0.25}(1620)^{0.75}$$ Substitute.

$$= 16{,}200$$ Use a calculator.

12.2 PARTIAL DERIVATIVES

In this section we extend the concept of derivative to functions of more than one variable, and study what are called *partial derivatives*.

First-Order Partial Derivatives

If $z = f(x, y)$ is a function of two variables, the first-order partial derivative of f with respect to x, written

$$f_x(x, y), \frac{\partial z}{\partial x}, \text{ or } f_x,$$

is found by treating y as a constant and taking the derivative of the function with respect to x. Similarly, the first-order partial derivative of f with respect to y, written

$$f_y(x, y), \frac{\partial z}{\partial y}, \text{ or } f_y,$$

is found by treating x as a constant and taking the derivative of the function with respect to y. Note that if $f(x, y) = x$, $f_x(x, y) = 1$ and

$f_y(x, y) = 0.$

EXAMPLE 12.4

If $z = f(x, y) = 3x^2 - 4xy + 1$, find f_x and f_y.

SOLUTION 12.4

To find f_x, treat y as a constant.

$f(x, y) = 3x^2 - 4xy + 1$ — Copy the function.

$f_x = 6x - 4y$ — Use the Power Rule on $3x^2$. Do *not* use the Product Rule on $-4xy$ since $-4y$ is treated as a constant.

$f_y = -4x$ — Both $3x^2$ and 1 are now treated as constants so each has a derivative of 0.

EXAMPLE 12.5

Find all the first-order partial derivatives for the following functions.

(a) $f(x, y) = \frac{x}{y}$

(b) $h(r, s) = e^{r+2s}$

(c) $h(u, v) = \sqrt{u^2 - 2v^2}$

(d) $f(x, y) = \frac{xy}{x^2 + y^2}$

SOLUTION 12.5

(a) $f(x, y) = \frac{x}{y}$ — Copy the function.

$f(x, y) = \frac{1}{y}x$ — Rewriting the function in this form may help.

$f_x = \frac{1}{y}$ — To find f_x, treat y as a constant.

$f(x, y) = xy^{-1}$ — Rewrite the function to apply the Power Rule.

$f_y = -xy^{-2}$	Treat x as a constant and apply the Power Rule.
(b) $h(r, s) = e^{r+2s}$	Copy the function.
$h_r = e^{r+2s}(1)$	Treat $2s$ as a constant. The derivative with respect to r of $r + 2s$ is 1.
$= e^{r+2s}$	Simplify.
$h_s = e^{r+2s}(2) = 2e^{r+2s}$	The derivative with respect to s of $r + 2s$ is 2.
(c) $h(u, v) = \sqrt{u^2 - 2v^2}$	Copy the function.
$= (u^2 - 2v^2)^{1/2}$	Rewrite the radical as a fractional exponent.
$h_u = \frac{1}{2}(u^2 - 2v^2)^{-1/2}(2u)$	Use the Chain Rule.
$= u(u^2 - 2v^2)^{-1/2}$	Simplify.
$= \dfrac{u}{\sqrt{u^2 - 2v^2}}$	Solution may be written in this form.
$h_v = \frac{1}{2}(u^2 - 2v^2)^{-1/2}(-4v)$	Treat u as a constant.
$= \dfrac{-2v}{\sqrt{u^2 - 2v^2}}$	Simplify.
(d) $f(x, y) = \dfrac{xy}{x^2 + y^2}$	Copy the function.
$f_x = \dfrac{(x^2 + y^2)y - (xy)(2x)}{(x^2 + y^2)^2}$	Use the Quotient Rule and treat y as a constant.
$= \dfrac{x^2y + y^3 - 2x^2y}{(x^2 + y^2)^2}$	Multiply out the numerator.
$= \dfrac{y^3 - x^2y}{(x^2 + y^2)^2}$	Combine like terms.

$$f_y = \frac{(x^2+y^2)x - xy(2y)}{(x^2+y^2)^2}$$ Use the Quotient Rule and treat x as a constant.

$$= \frac{x^3 + xy^2 - 2xy^2}{(x^2+y^2)^2}$$ Multiply out the numerator.

$$= \frac{x^3 - xy^2}{(x^2+y^2)^2}$$ Combine like terms.

Evaluating Partial Derivatives

Recall that if $f(x) = x^2 + 1$, $f(2)$ indicates that x should be replaced with 2. Similarly, if $f_x(x, y) = 6x - 4y$, $f_x(1, 2)$ means replace x with 1, and y with 2 in the first-order partial derivative taken with respect to x.

$$f_x(1, 2) = 6(1) - 4(2) = -2$$

EXAMPLE 12.6

For $z = f(x, y) = 4x^3y^2 + 3$, find

(a) $f_x(1, -2)$

(b) $f_y(2, 3)$

SOLUTION 12.6

(a) First find f_x by treating y as a constant:

$$f_x = 12x^2y^2$$

Now evaluate f_x with $x = 1$ and $y = -2$:

$$f_x(1, -2) = 12(1)^2(-2)^2 = 48$$

(b) Find f_y by treating x as a constant:

$$f_y = 8x^3y$$

Evaluate $f_y(2, 3)$:

$$f_y(2, 3) = 8(2)^3(3) = 192$$ Substitute $x = 2$, $y = 3$.

Applications Using Partial Derivatives

Recall from chapter 10 that we used the first derivative to find instantaneous rates of change. Partial derivatives can also be interpreted as instantaneous rates of change. The partial derivative of f with respect to x at (x_0, y_0), $f_x(x_0, y_0)$, is the instantaneous rate of change of f in the x direction when y is held constant. The partial derivative of f with respect to y at (x_0, y_0), $f_y(x_0, y_0)$, is the instantaneous rate of change of f in the y direction when x is held constant. We can also interpret the partial derivative as the approximate change in f per unit change in one variable when the other variable is held constant.

EXAMPLE 12.7

For the Cobb-Douglas production formula from Example 12.3,

$$f(x, y) = 30x^{0.25}y^{0.75}$$

find f_x, called the *marginal productivity of labor*, and f_y, called the *marginal productivity of capital*. If the calculator company currently uses 20 units of labor and 1620 units of capital, find the marginal productivity of labor and the marginal productivity of capital.

SOLUTION 12.7

$f(x, y) = 30x^{0.25}y^{0.75}$ — Copy the function.

$f_x(x, y) = 30(0.25)x^{0.25-1}y^{0.75}$ — Use the Power Rule and treat y as a constant.

$= 7.5x^{-0.75}y^{0.75}$ — Simplify.

$f_y(x, y) = 30x^{0.25}(0.75)y^{0.75-1}$ — Use the Power Rule and treat x as a constant.

$= 22.5x^{0.25}y^{-0.25}$ — Simplify.

To find the marginal productivity of labor and the marginal productivity of capital, find $f_x(20, 1620)$ and $f_y(20, 1620)$.

$f_x(20, 1620) = 7.5(20)^{-0.75}(1620)^{0.75}$ — Substitute $x = 20$ and $y = 1620$ in f_x.

≈ 202 — Use a calculator.

$f_y(20, 1620) = 22.5(20)^{0.25}(1620)^{-0.25}$ Substitute $x = 20$ and $y = 1620$ in f_y.

≈ 7.5 Use a calculator.

At the current levels of labor and capital, each unit increase in labor (keeping capital constant) will result in an increase of production by approximately 202 calculators. At the current levels of labor and capital, an increase of one unit of capital (keeping labor constant) will result in an increase of production of approximately 7.5 calculators. Assuming similar costs for labor and capital the management should consider increasing labor, if their goal is to increase production.

Higher-Order Partial Derivatives

Recall from section 9.2, we can find the second derivative by taking the derivative of the first derivative. We can find higher-order partial derivatives in a similar manner. Note, however, that a function of two variables will have *four* second-order partial derivatives:

f_{xx} means take the partial derivative of f_x with respect to x.

f_{xy} means take the partial derivative of f_x with respect to y.

f_{yy} means take the partial derivative of f_y with respect to y.

f_{yx} means take the partial derivative of f_y with respect to x.

The notation for second-order partial derivatives for $z = f(x, y)$ can be written several ways:

$$f_{xx} = f_{xx}(x, y) = \frac{\partial^2 z}{\partial x^2} = \frac{\partial^2 f}{\partial x^2}$$

$$f_{xy} = f_{xy}(x, y) = \frac{\partial^2 z}{\partial y \partial x} = \frac{\partial^2 f}{\partial y \partial x}$$

$$f_{yy} = f_{yy}(x, y) = \frac{\partial^2 z}{\partial y^2} = \frac{\partial^2 f}{\partial y^2}$$

$$f_{yx} = f_{yx}(x, y) = \frac{\partial^2 z}{\partial x \partial y} = \frac{\partial^2 f}{\partial x \partial y}$$

EXAMPLE 12.8

Find $f_{xx}(x, y)$, $f_{xy}(x, y)$, $f_{yx}(x, y)$, and $f_{yy}(x, y)$ for each function.

(a) $f(x, y) = 4x^2y^2 + 6x + 2y$

(b) $f(x, y) = \ln x^2y^2$

SOLUTION 12.8

(a) $f(x, y) = 4x^2y^2 + 6x + 2y$ — Copy the function.

$f_x(x, y) = 8xy^2 + 6$ — Find the first-order partial derivatives.

$f_y(x, y) = 8x^2y + 2$

$f_{xx}(x, y) = 8y^2$ — Use f_x to find f_{xx} and f_{xy}.

$f_{xy}(x, y) = 16xy$

$f_{yx}(x, y) = 16xy$ — Use f_y to find f_{yx} and f_{yy}.

$f_{yy}(x, y) = 8x^2$

(b) $f(x, y) = \ln x^2y^2$ — Copy the function.

$f_x(x, y) = \frac{1}{x^2y^2} \cdot 2xy^2$ — Find the first-order partial derivatives.

$= \frac{2}{x} = 2x^{-1}$ — Simplify.

$f_y(x, y) = \frac{1}{x^2y^2} \cdot 2x^2y$

$= \frac{2}{y} = 2y^{-1}$ — Simplify.

$f_{xx}(x, y) = -2x^{-2}$ — Use f_x to find f_{xx} and f_{xy}.

$f_{xy}(x, y) = 0$ — Since $2x^{-1}$ is a constant, the derivative equals 0.

$f_{yx}(x, y) = 0$ — Since $2y^{-1}$ is a constant, the derivative is 0.

$f_{yy}(x, y) = -2y^{-2}$ Use the Power Rule.

Note: In Example 12.8 (a) and (b), $f_{xy} = f_{yx}$. When the given function and its partial derivatives meet certain conditions, this will be the case. Most of the functions you will encounter will meet these conditions.

12.3 LOCAL EXTREMA

Just as we were able to extend the concept of derivatives to functions of two variables, we can extend the concept of local extrema to functions of two variables. In this section we will find critical points and use the Second Partial Derivative Test to find local maximums and minimums for surfaces.

Critical Points

Recall that if $f(x)$ had a local maximum or minimum, it occurred where $f'(c) = 0$ or where $f'(c)$ was undefined. Similarly, if $f(x, y)$ has a local maximum or minimum at (x_0, y_0), then $f_x(x_0, y_0) = 0$ and $f_y(x_0, y_0) = 0$. We then call (x_0, y_0) a *critical point* of *f*. We can find critical points using the following steps.

1. Find f_x and f_y.
2. Set $f_x = 0$ and $f_y = 0$ and solve. This step may require solving a system of equations.

EXAMPLE 12.9

Find the critical points for each function.

(a) $f(x, y) = x^2 - 4x + 2y^2 + 6y + 2$

(b) $f(x, y) = -x^2 + 4x - y^2 - 6y$

(c) $f(x, y) = y^2 + 4y - x^2 + 8x$

SOLUTION 12.9

(a) $f(x, y) = x^2 - 4x + 2y^2 + 6y + 2$ Copy the function.

$f_x(x, y) = 2x - 4 \quad f_y(x, y) = 4y + 6$

Find f_x and f_y.

$2x - 4 = 0 \quad 4y + 6 = 0$

$x = 2 \quad y = -3/2$

Set the partial derivatives equal to 0 and solve.

$(2, -3/2)$ is a critical point.

(b) $f(x, y) = -x^2 + 4x - y^2 - 6y$ Copy the function.

$f_x(x, y) = -2x + 4 \quad f_y(x, y) = -2y - 6$

Find f_x and f_y.

$-2x + 4 = 0 \quad -2y - 6 = 0$

$x = 2 \quad y = -3$

Set the partial derivatives equal to 0 and solve.

$(2, -3)$ is a critical point.

(c) $f(x, y) = y^2 + 4y - x^2 + 8x$ Copy the function.

$f_x(x, y) = -2x + 8 \quad f_y(x, y) = 2y + 4$

Find f_x and f_y.

$-2x + 8 = 0 \quad 2y + 4 = 0$

$x = 4 \quad y = -2$

Set the partial derivatives equal to 0 and solve.

$(4, -2)$ is a critical point.

Local Extrema

The test for local extrema of functions of two variables is similar to the Second Derivative Test for functions of one variable. As with functions of one variable, a critical point is not necessarily a local maximum or minimum. However, all local extrema occur at critical points. The Second Partial Derivative Test not only includes the criteria for local extrema, but also for *saddle points*, points that are local maximums when planes intersect the curve in one direction, but are local minimums when planes intersect the curve in another direction. The conditions for the Second Partial Derivative test can be found in any advanced calculus textbook. We present here the essential steps to follow to find local maximums and minimums, and saddle points.

Using The Second Partial Derivative Test for Local Extrema

1. Find critical point(s) (x_0, y_0) by setting f_x and f_y equal to 0 and

solving.

2. Find $D = f_{xx}f_{yy} - f_{xy}^{\ 2}$ where $f_{xy}^{\ 2} = (f_{xy})^2$.
3. If $D(x_0, y_0) > 0$ and $f_{xx}(x_0, y_0) > 0$, then f has a local minimum at (x_0, y_0).
4. If $D(x_0, y_0) > 0$ and $f_{xx}(x_0, y_0) < 0$, then f has a local maximum at (x_0, y_0).
5. If $D(x_0, y_0) < 0$ then f has a saddle point at (x_0, y_0).
6. If $D(x_0, y_0) = 0$, the test fails.

EXAMPLE 12.10

Find the local extrema for each function.

(a) $f(x, y) = x^2 - 4x + 2y^2 + 6y + 2$

(b) $f(x, y) = -x^2 + 4x - y^2 - 6y$

(c) $f(x, y) = y^2 + 4y - x^2 + 8x$

SOLUTION 12.10

We have already found the critical points for each function in Example 12.9. We proceed from that point in the solution.

(a) $f(x, y) = x^2 - 4x + 2y^2 + 6y + 2$ has a critical point at (2, –3/2).

$f_x = 2x - 4 \quad f_y = 4y + 6$ First-order partial derivatives.

$f_{xx} = 2 \quad f_{yy} = 4$ Find f_{xx}, f_{yy}, and f_{xy}.

$f_{xy} = 0$

$D\left(2, -\frac{3}{2}\right) = 2(4) - 0^2 = 8$ Find $D = f_{xx}f_{yy} - f_{xy}^{\ 2}$.

Since $D > 0$ and $f_{xx} > 0$, (2, –3/2) is a local minimum.

(b) $f(x, y) = -x^2 + 4x - y^2 - 6y$ has a critical point at (2, –3).

$f_x = -2x + 4 \quad f_y = -2y - 6$ First-order partial derivatives.

$f_{xx} = -2 \quad f_{yy} = -2$ Find f_{xx}, f_{yy}, and f_{xy}.

$f_{xy} = 0$

$D(2, -3) = -2(-2) - 0^2 = 4$ Find $D = f_{xx}f_{yy} - f_{xy}^2$.

Since $D > 0$ and $f_{xx} < 0$, $(2, -3)$ is a local maximum.

(c) $f(x, y) = y^2 + 4y - x^2 + 8x$ has a critical point at $(4, -2)$.

$f_x = -2x + 8$ $f_y = 2y + 4$ First-order partial derivatives.

$f_{xx} = -2$ $f_{yy} = 2$ Find f_{xx}, f_{yy}, and f_{xy}.

$f_{xy} = 0$

$D(4, -2) = -2(2) - 0^2 = -4$ Find D.

Since $D < 0$, $(4, -2)$ is a saddle point.

12.4 LAGRANGE MULTIPLIERS

In this section we will solve problems in restricted regions that take the form

$$\begin{gathered}\text{Maximize or Minimize } f(x, y)\\ \text{Subject to } g(x, y) = 0\end{gathered}$$

where the function $g(x, y)$ acts as a restriction on the domain (x, y) for f and is called a constraint. The technique we will use to solve this type of max-min problem is called the method of Lagrange multipliers. The steps for using the method of *Lagrange multipliers* follow.

1. Write the function to be maximized or minimized in the form $z = f(x, y)$. Write the constraint in the form $g(x, y) = 0$.
2. Form a new function F, where $F(x, y, z) = f(x, y) + \lambda g(x, y)$ and λ is a parameter called the *Lagrange multiplier.*
3. Find the partial derivatives of F with respect to x, then y, then λ, set each partial derivative equal to 0 and solve. The solution(s) is called a critical point(s).
4. Evaluate f using the values found in step 3. The largest z value found will be the maximum and the smallest z value will be the minimum.

EXAMPLE 12.11

Maximize $f(x, y) = xy$ subject to $x + y = 4$.

SOLUTION 12.11

$f(x, y) = xy$	The function to be maximized is already in the correct form.
$x + y = 4 \quad \Rightarrow \quad x + y - 4 = 0$	Set the constraint equal to 0.
$F(x, y, z) = xy + \lambda(x + y - 4)$	Form F.
$F(x, y, z) = xy + \lambda x + \lambda y - 4\lambda$	Simplify F.
$F_x = y + \lambda \quad F_y = x + \lambda \quad F_\lambda = x + y - 4$	Find each partial derivative.
Solve: (1) $y + \lambda = 0$ (2) $x + \lambda = 0$ (3) $x + y - 4 = 0$	Set each partial derivative equal to 0 and solve the resulting system of equations.
(1) $y = -\lambda$ (2) $x = -\lambda$ (3) $-\lambda - \lambda - 4 = 0$ $-2\lambda = 4$ $\lambda = -2$	Solve (1) and (2) for x and y and substitute into (3).

If $\lambda = -2$, $x = -(-2) = 2$, and $y = -(-2) = 2$.
Evaluate $f(2, 2) = 2(2) = 4$.
Therefore, 4 is the maximum value of f.

EXAMPLE 12.12

The Cobb-Douglas production function for a company that produces calculators is

$$f(x, y) = 30x^{0.25}y^{0.75}$$

where x is the number of units of labor and y is the number of units of capital required to produce $f(x, y)$ calculators. Each unit of labor costs \$20 and each unit of capital costs \$1620. If the company has budgeted \$10,000 for production, how should this money be allocated to maximize production?

SOLUTION 12.12

Maximize $f(x, y) = 30x^{0.25}y^{0.75}$ — This is the function to be maximized.

$20x + 1620y = 10000$ — The constraint function contains the restriction that costs must equal the budgeted amount of \$10,000.

$20x + 1620y - 10000 = 0$ — Set the constraint equal to 0.

$F = 30x^{0.25}y^{0.75} + \lambda(20x + 1620y - 10000)$ — Form F.

$F = 30x^{0.25}y^{0.75} + 20\lambda x + 1620\lambda y - 10000\lambda$ — Simplify F.

$F_x = 7.5x^{-0.75}y^{0.75} + 20\lambda$ — Find each partial derivative.

$F_y = 22.5x^{0.25}y^{-0.25} + 1620\lambda$

$F_\lambda = 20x + 1620y - 10000$

Solve:

(1) $7.5x^{-0.75}y^{0.75} + 20\lambda = 0$

(2) $22.5x^{0.25}y^{-0.25} + 1620\lambda = 0$

(3) $20x + 1620y - 10000 = 0$

Set each partial derivative equal to 0 and solve the resulting system of equations.

Solve equations (1) and (2) for λ and set them equal to each other:

$-\frac{7.5}{20}x^{-0.75}y^{0.75} = -\frac{22.5}{1620}x^{0.25}y^{-0.25}$ — Multiply by $x^{0.75}y^{0.25}$.

$-\frac{7.5}{20}y = -\frac{22.5}{1620}x$

$y = \frac{1}{27}x$ — Solve for y.

Substitute for y in equation (3):

$20x + 1620\left(\frac{1}{27}x\right) - 10000 = 0$ — Simplify and solve for x.

$20x + 60x - 10000 = 0$

$80x = 10000$

$x = 125$

$y = \frac{1}{27}(125) \approx 4.6$

Then the maximum $f(125, 4.6) = 30(125)^{0.25}(4.6)^{0.75} = 315$ units

subject to the constraint that only $10,000 be spent for production.

12.5 ITERATED INTEGRALS

In this section, we continue to extend our concepts for functions of one variable to functions of more than one variable. In particular, in this section we will extend integration to functions of two or more variables.

Just as we treat one variable as a constant to perform partial differentiation, we treat one variable as a constant to perform integration. For example, the notation $\int f(x, y)\,dx$ would indicate that x is to be treated as a variable and y is to be treated as a constant. Similarly $\int f(x, y)\,dy$ indicates that y is a variable and x is a constant.

EXAMPLE 12.13

Evaluate.

(a) $\int (4x^2y + 6x)\,dx$

(b) $\int (4x^2y + 6x)\,dy$

SOLUTION 12.13

(a) $\int (4x^2y + 6x)\,dx$	Copy the integral.
$= \int 4x^2ydx + \int 6xdx$	The integral of a sum is the sum of the integrals.
$= 4y\int x^2dx + 6\int xdx$	Since y is treated as a constant, it can be factored out of the integral.
$= 4y\left(\frac{x^3}{3}\right) + 6\left(\frac{x^2}{2}\right) + C$	Use the Power Rule.
$= \frac{4}{3}x^3y + 3x^2 + C$	Simplify.

(b) $\int (4x^2y + 6x)\, dy$ — Copy the integral.

$= \int 4x^2y\,dy + \int 6x\,dy$ — The integral of a sum is the sum of the integrals.

$= 4x^2 \int y\,dy + 6x \int dy$ — Since x is treated as a constant, it can be factored out of the integral.

$= 4x^2\left(\frac{y^2}{2}\right) + 6xy + C$ — Use the Power Rule.

$= 2x^2y^2 + 6xy + C$ — Simplify.

Definite Integrals

The dx in $\int f(x, y)\, dx$ indicates that x is treated as a variable in the indefinite integral. In a definite integral, it also implies that the limits of integration are x values. Similarly $\int_a^b f(x, y)\, dy$ implies that a and b are to be substituted for y after the integration.

EXAMPLE 12.14

Evaluate.

(a) $\int_0^1 (4x^2y + 6x)\, dx$

(b) $\int_0^1 (4x^2y + 6x)\, dy$

SOLUTION 12.14

We have already integrated these functions in the previous example. We proceed from that point.

(a) $\int_0^1 (4x^2y + 6x)\, dx$ — dx indicates $x = 0$ and $x = 1$ are the limits.

$= \left(\frac{4}{3}x^3y + 3x^2\right)\Big|_0^1$

$= \left[\frac{4}{3}(1)^3y + 3(1)^2\right] - \left[\frac{4}{3}(0)^3y + 3(0)^2\right]$ — Use the Fundamental Theorem to substitute $x = 1$ and $x = 0$ and subtract.

$= \frac{4}{3}y + 3 - 0 = \frac{4}{3}y + 3$ — Simplify.

(b) $\int_0^1 (4x^2y + 6x)\,dy$ — dy indicates $y = 0$ and $y = 1$ are the limits.

$$= (2x^2y^2 + 6xy)\Big|_0^1$$

$$= [2x^2(1)^2 + 6x(1)] - [2x^2(0)^2 + 6x(0)]$$

Use the Fundamental Theorem to substitute $y = 1$ and $y = 0$ and subtract.

$$= 2x^2 + 6x - 0 = 2x^2 + 6x$$

Simplify.

Iterated Integrals

Note in the previous example that $\int_a^b f(x, y)\,dx$ resulted in a function of one variable, y. Also $\int_c^d f(x, y)\,dy$ resulted in a function of one variable, x. We can now proceed to integrate each of these functions with respect to the remaining variable.

EXAMPLE 12.15

Evaluate.

$$\int_2^3 \left[\int_0^1 (4x^2y + 6x)\,dx\right] dy$$

SOLUTION 12.15

We already found that $\int_0^1 (4x^2y + 6x)\,dx = \frac{4}{3}y + 3$.

Now,

$$\int_2^3 \left(\frac{4}{3}y + 3\right)dy = \int_2^3 \frac{4}{3}y\,dy + \int_2^3 3\,dy$$

The integral of a sum is the sum of the integrals.

$$= \frac{4}{3}\left(\frac{y^2}{2}\right)\Bigg|_2^3 + 3y\Big|_2^3$$

Use the Fundamental Theorem.

$$= \frac{2}{3}(3^2 - 2^2) + 3(3 - 2)$$

Evaluate.

$$= \frac{2}{3}(5) + 3(1) = \frac{19}{3}$$

Simplify.

The process we've just completed, integrating once with respect to one variable and then integrating with respect to the other variable, is called double integration. The expression $\iint f(x, y)\,dxdy$ is also referred to as an iterated integral.

Geometric Interpretation of the Double Integral

We now extend the geometric interpretation of a single integral. Recall that $\int_a^b f(x)\,dx$ represented the area bounded by $f(x)$, the x-axis and the lines $x = a$ and $x = b$. In an analgous manner, we note that $\int_c^d \int_a^b f(x, y)\,dxdy$ can be interpreted as the volume of the solid bounded above by $z = f(x, y)$, below by $z = 0$, and has sides formed by the planes $x = a$, $x = b$, $y = c$, and $y = d$. Note also that the sides of the solid form a rectangle in the $z = 0$ plane where $a \le x \le b$ and $c \le y \le d$.

EXAMPLE 12.16

Determine the volume of the solid bounded by $f(x, y) = 3x + y$ over the rectangle $\{(x, y) \mid 2 \le x \le 4 \text{ and } 1 \le y \le 3\}$.

SOLUTION 12.16a

$$\int_1^3 \int_2^4 (3x + y)\,dxdy$$

Set up the required double integral. Since dx is listed first, the inner limits of integration are 2 and 4.

$$\int_2^4 (3x + y)\,dx = \left(\frac{3x^2}{2} + xy\right)\Bigg|_2^4$$

Integrate with respect to x, treating y as a constant.

$$= \left[\frac{3(4)^2}{2} + 4(y)\right] - \left[\frac{3(2)^2}{2} + 2(y)\right]$$

Substitute $x = 4$ and $x = 2$ and subtract.

$$= (24 + 4y) - (6 + 2y)$$

Simplify.

$$= 18 + 2y$$

Now integrate with respect to y:

$$\int_1^3 (18 + 2y)\,dy = (18y + y^2)\Big|_1^3$$

$$= [18(3) + (3)^2] - [18(1) + (1)^2]$$

Substitute $y = 3$ and $y = 1$ and subtract.

$$= (54 + 9) - (18 + 1) = 44$$

Simplify.

SOLUTION 12.16b

Let's rework the given problem reversing the order of integration. That is, let's find

$\int_2^4 \int_1^3 (3x+y)\, dydx$

First evaluate:

$\int_1^3 (3x+y)\, dy = \left(3xy + \frac{y^2}{2}\right)\Big|_1^3$ Integrate with respect to y, treating x as a constant.

$= \left[3x(3) + \frac{(3)^2}{2}\right] - \left[3x(1) + \frac{(1)^2}{2}\right]$ Substitute $y = 3$ and $y = 1$ and subtract.

$= \left(9x + \frac{9}{2}\right) - \left(3x + \frac{1}{2}\right)$ Simplify.

$= 6x + 4$

Now integrate with respect to x:

$\int_2^4 (6x+4)\, dx = (3x^2 + 4x)\Big|_2^4$

$= [3(4)^2 + 4(4)] - [3(2)^2 + 4(2)]$ Substitute $x = 4$ and $x = 2$ and subtract.

$= 64 - 20 = 44$ Simplify.

The point that these two solutions emphasize is that the volume of a solid bounded above by $z = f(x, y)$ and in the $z = 0$ plane by the rectangle $\{(x, y) \mid a \le x \le b \text{ and } c \le y \le d\}$ can be evaluated by the use of either iterated integral. That is

$$\int_c^d \int_a^b f(x, y)\, dxdy = \int_a^b \int_c^d f(x, y)\, dydx.$$

12.6 METHOD OF LEAST SQUARES

In this section we will use formulas that result from partial differentiation to find a line that best describes a set of data. The method for finding this line is called *Least Squares Approximation* or the *Method of Least Squares*. The formulas are derived in most textbooks—here we will only make use of those results.

Scatter Diagrams

In chapter 6 we presented several ways of displaying data. Another method, called a **scatter diagram**, plots points in an x-y coordinate system.

EXAMPLE 12.17

Plot a scatter diagram for the given data points.
(a) (1, 6), (2, 7), (3, 11), (4, 13)
(b) (1, 4), (2, 4), (3, 2), (4, 1)

SOLUTION 12.17

(a)

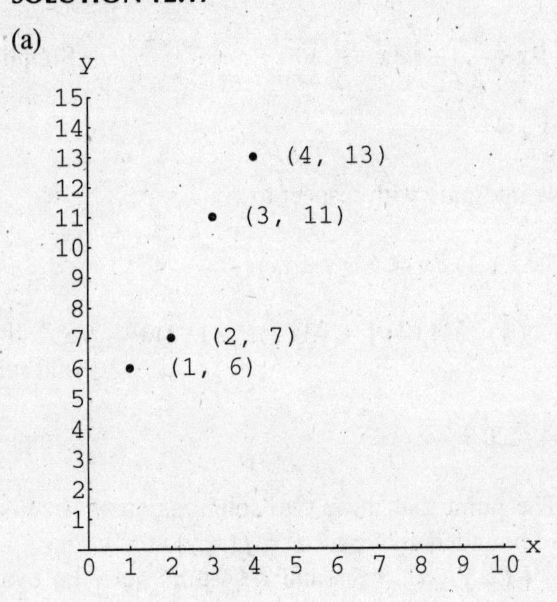

(b)

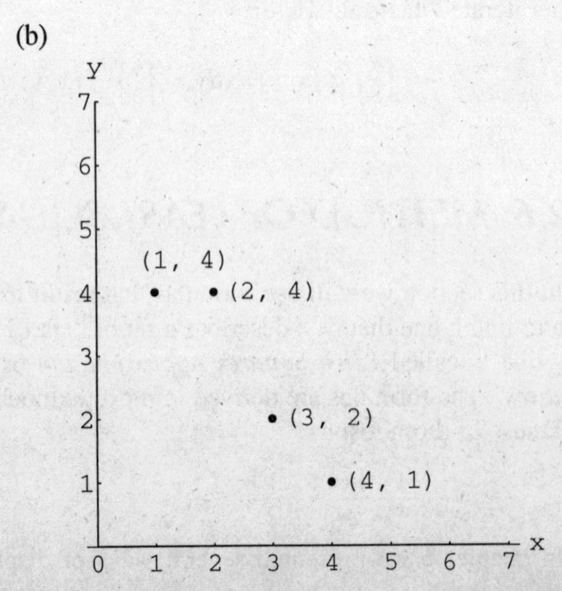

Least Squares Line

Our goal is to find a line that approximates our data. Return to the scatter diagram from Example 12.17 (a) and try to draw a line that approximates the relationship:

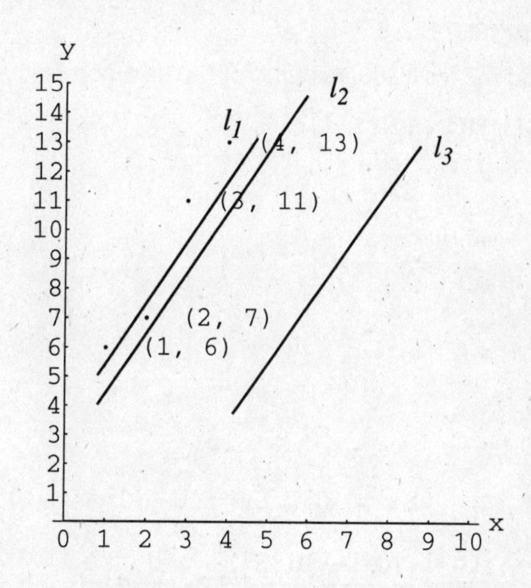

We've drawn several lines, some of which do a better job of describing this data than others. For example, l_3 doesn't contain any of the data points and would not be a good choice to represent our data.

A good choice for a line that represents our data is one where the distances from the data points to the line is minimized in some sense. It can be shown that the following technique yields such a line.

The Equation of the Least Squares Line
For data points (x_1, y_1), (x_2, y_2), ... (x_n, y_n), the equation of the least squares line is $y = mx + b$ where

$$m = \frac{n\sum xy - \sum x \sum y}{n\sum x^2 - (\sum x)^2}$$

$$b = \frac{\sum y - m\sum x}{n}$$

Note that n equals the number of data points, Σ is the symbol for summation, m is the slope of the line, and b is the y-intercept of the line.

One technique for organizing your data to efficiently calculate m and b is to list in columns the x-coordinates, y-coordinates, the product of each x-coordinate times its corresponding y-coordinate, and the x-coordinates squared. Then sum each column. Study the following example.

EXAMPLE 12.18

Find the least squares line for the data points (1, 6), (2, 7), (3, 11), (4, 13).

SOLUTION 12.18

x	y	xy	x^2
1	6	6	1
2	7	14	4
3	11	33	9
4	13	52	16
$\Sigma x = 10$	$\Sigma y = 37$	$\Sigma xy = 105$	$\Sigma x^2 = 30$

$$m = \frac{4(105) - 10(37)}{4(30) - (10)^2} = \frac{50}{20} = \frac{5}{2}$$

$$b = \frac{37 - \frac{5}{2}(10)}{4} = \frac{12}{4} = 3$$

Therefore, the equation of the least squares line is $y = \frac{5}{2}x + 3$.

Recall that these were the data points from Example 12.17 (a). Here is that scatter diagram and the graph of the least squares line:

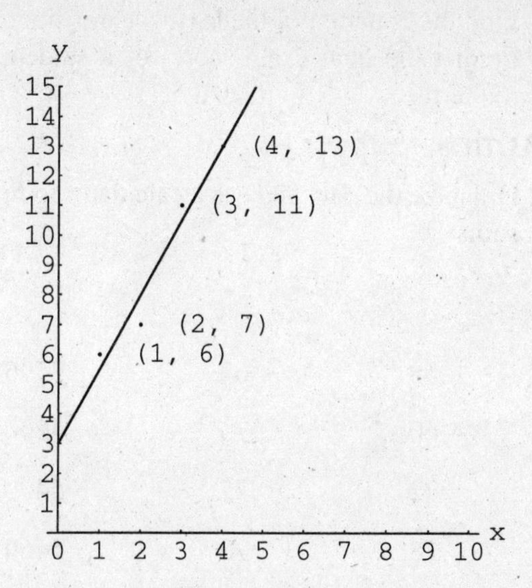

The use of a calculator will make the summing and squaring process easier. However, for more complicated applications, there are computer programs available that allow you to enter the data points and have the program provide the rest—the equation, and the graph!

Applications

Once we find the equation of the least squares line, we can use that equation to approximate a particular y-value given a specific x-value. Read the following example.

EXAMPLE 12.19

A group of enterprising students in Dr. Smith's class collected data from the previous semester for the midterm and final exam scores for 8 students. They collected the following:

Midterm	Final
58	62
60	70
72	75
78	70

Midterm	Final
82	79
85	89
92	90
95	95

(a) Find the equation of the least squares line.
(b) Predict the final exam score for a student who scored an 88 on the midterm.

SOLUTION 12.19

(a) Organize the data and use a calculator to find the squares and columns sums:

x	y	xy	x^2
58	62	3596	3364
60	70	4200	3600
72	75	5400	5184
78	70	5460	6084
82	79	6478	6724
85	89	7565	7225
92	90	8280	8464
95	95	9025	9025
$\Sigma x = 622$	$\Sigma y = 630$	$\Sigma xy = 50004$	$\Sigma x^2 = 49670$

$$m = \frac{8(50004) - 622(630)}{8(49670) - (622)^2} = \frac{8172}{10476} = 0.78$$

$$b = \frac{630 - 0.78(622)}{8} = 18.11.$$

Therefore, the equation of the least squares line is $y = 0.78x + 18$.

(b) To predict the final exam score for a student who scored an 88 on the midterm, find y when $x = 88$:
$y = 0.78(88) + 18 = 87$

Practice Exercises

1. Evaluate the given function at the indicated value.

   (a) $f(x, y) = 2x + 3\sqrt{y}$ at (3, 4)

   (b) $f(x, y) = x^2 - 4xy + y^2$ at (1, 2)

   (c) $f(x, y, z) = 2xy + 3yz + z^2$ at (0, 1, –1)

2. Find all the first-order partial derivatives for the following functions.

   (a) $f(x, y) = \dfrac{x+1}{y-2}$

   (b) $f(x, y) = 2e^{3x^2 - y}$

   (c) $h(u, v) = \dfrac{\sqrt{u+2v}}{uv^2}$

3. For $z = 6x^2y^3 + 4y$, find

   (a) $f_x(2, -1)$

   (b) $f_y(2, -1)$

   (c) $f_x(0, 3)$

   (d) $f_y(0, 3)$

4. Find $f_{xx}(x, y)$, $f_{xy}(x, y)$, $f_{yx}(x, y)$ and $f_{yy}(x, y)$ for each function.

   (a) $f(x, y) = 3xy^3 + 4x - 5y$

   (b) $f(x, y) = e^{x^2y^2}$

5. Find the local extrema for each function.

   (a) $f(x, y) = -2x^2 + 6x - y^2 + 4y$

   (b) $f(x, y) = 4x^2 + 8x + y^2 + 6y + 8$

   (c) $f(x, y) = 4x^2 - 6x - 2y^2 + 5y + 3$

6. Evaluate.

   (a) $\int (x^3y^2 + 4x)\,dx$

   (b) $\int (x^3y^2 + 4x)\,dy$

   (c) $\int_0^1 (x^3y^2 + 4x)\,dx$

   (d) $\int_0^1 (x^3y^2 + 4x)\,dy$

7. Evaluate.

   (a) $\int_1^2\int_0^1 (x^3y^2 + 4x)\,dxdy$

   (b) $\int_0^1\int_1^2 (x^3y^2 + 4x)\,dydx$

8. Find the least squares line for the data points (1, 6), (2, 4), (3, 3), (4, 1).

Answers

1. (a) 12

   (b) –3

   (c) –2

2. (a) $f_x = \frac{1}{y-2}$

   $f_y = -\frac{x+1}{(y-2)^2}$

   (b) $f_x = 12xe^{3x^2-y}$

   $f_y = -2e^{3x^2-y}$

   (c) $f_u = -\frac{1}{u^2v\sqrt{u+2v}}$

   $f_v = -\frac{3v+2u}{uv^3\sqrt{u+2v}}$

3. (a) –24

   (b) 76

   (c) 0

   (d) 4

4. (a) $f_{xx} = 0$

   $f_{xy} = 9y^2$

   $f_{yx} = 9y^2$

   $f_{yy} = 27xy^2$

   (b) $f_{xx} = 2e^{x^2y^2}(1+2x^2)$

   $f_{xy} = 4xye^{x^2y^2}$

   $f_{yx} = 4xye^{x^2y^2}$

   $f_{yy} = 2e^{x^2y^2}(1+2y^2)$

5. (a) (3/2, 2) local maximum

   (b) (–1, –3) local minimum

   (c) $\left(\frac{3}{4}, \frac{5}{4}\right)$ saddle point

6. (a) $\frac{1}{2}x^4y^2 + 2x^2 + C$

   (b) $\frac{1}{3}x^3y^3 + 4xy + C$

   (c) $\frac{1}{4}y^2 + 2$

   (d) $\frac{1}{3}x^3 + 4x$

7. (a) 31/12

   (b) 31/12

8. $y = -\frac{8}{5}x + \frac{15}{2}$

Index

A

B

C

D

E

F

G

H

I

S

T

U

V

W

X

Y

Z